Reflections on Volcanic Glass

Proceedings of the 2021 International
Obsidian Conference

Edited by Lucas R. M. Johnson,
Kyle P. Freund, and Nicholas Tripcevich

Foreword by M. Steven Shackley

© 2024 Regents of the University of California
Published by eScholarship, Berkeley, CA
1st edition
PDF: 978-0-9982460-7-9
POD: 978-0-9982460-8-6

Available open access at: https://escholarship.org/uc/item/75c689n2

Publication of this book was made possible by funding from
UC Berkeley Library and Berkeley Research Impact Initiative Funds.

Cover image courtesy of Proyecto Arqueológico Conchopata (PAC) (PI Wm. Isbell and A. Cook)
Production: Westwood Press, www.westwoodpress.com
Design: Morgane Leoni

Contents

List of figures

List of tables

FOREWORD

M. Steven Shackley

I admit I was honored to be the keynote speaker for the 2021 International Obsidian Conference (IOC) from which these proceedings come. I am personal friends and colleagues with many of the participants to this day and thus have "skin in the game," so to speak. The conference, as the editors outline, was to be held at the University of California, Berkeley, where I taught geoarchaeology in the departments of Anthropology and Earth & Planetary Science for 23 years. We all looked forward to being together, as friends and colleagues. Unfortunately, the COVID-19 pandemic made that impossible. There are, however, certain advantages to a Zoom conference, and we were able to interface with people who have an interest in obsidian and provenance studies worldwide—something that would not have been possible with an in-person meeting in California.

At least 70 people participated in the IOC, with some of the European participants staying through the night to give their lectures and critique. I believe it was an educational experience for all; certainly, for me, it was. I wish, as I am sure the editors do, that more conference participants would have submitted papers, but such is always the case given time commitments.

Our small sub-discipline has changed substantially in the nearly 40 years I've been involved, and especially in the last decade or so, and it is thus imperative that we keep in touch. I notice in many chapters that the authors tend to cite mostly others working in their region, sometimes only citing the more recent work—I have been guilty of this as well. Proceedings like this may help mitigate that issue. Another larger volume on a similar subject, which is about to be submitted to Springer, may aid in this cause too (Le Bourdonnec, et al. 2024). In fact, many of the conference participants are also authors in that volume, which could be one reason that the proceedings did not capture as many papers as might have been possible.

When obsidian provenance studies really began to become a major part of the archaeological endeavor, there were only a handful of us, particularly in the US, who handled all the work, which was growing at a rapid pace in a rapidly growing discipline. A short history here is relevant (see Shackley 2011 for more depth). For good or bad, the nexus was at, or stemming from, the UC Berkeley campus and the Lawrence Berkeley National Laboratory up the hill. Most involved and constantly sharing data were the following: Craig Skinner in Oregon; Richard Hughes in Sacramento (then in the Bay Area); Tom Jackson in Aptos, California; Kathy Davis, the brightest of us all, in Davis, California; Paul Bouey at University of California, Davis; Mike Glascock at University of

Missouri; Fred Nelson at Brigham Young University; and me. These data were all based on laboratory X-ray fluorescence (XRF) and Neutron Activation Analysis (NAA) at the University of Missouri Research Reactor (MURR) center and all calibrated using international standards. This was crucial for sharing data, something apparently more difficult in the portable XRF (pXRF) world, which I have lamented perhaps at too much length in the past. At that time, it was possible to send any of our data (not artifacts), from our own instruments, to someone else, and since we all calibrated with the same standards, even though comparing XRF and NAA (see Glascock 2011), we could determine the source with some degree of confidence. And we could always normalize another lab's data through analysis of the same international standards to our own instruments and determine probable source. It has been said that this is more difficult, or impossible, to do with pXRF, although that is not my experience, as is obvious in this volume. Some in that "old group" have become somewhat concerned about the silo science of individual pXRF laboratories, which cannot share data, since few analyze a known standard that allows for normalization between instruments. Without that ability, I fail to see the point in sharing data. Again, however, this has not been my experience with pXRF labs at Santa Clara University, Far Western Anthropological Research Group in Nevada, the Gila River Indian Community in Arizona, Southern Methodist University, or the University of Edinburgh, nor in the chapters in this volume. To be fair, I think this is improving; hopefully, this is not just because of my complaining but also because of the realization that if we cannot share data, then it may all fall apart. Having said that, I find these proceedings very hopeful and somewhat beyond what I could have imagined during my own work in the 1980s and 1990s (c.f. Shackley 1998).

I believe the reader will find of interest the fact that not all the papers here are focused on provenance or the XRF method. Many papers find that meshing the technological analysis of obsidian artifacts—including utilized flakes as well as formal tools (or obsidian jewelry)—can address issues that provenance alone can never achieve. Obsidian, as we all know, has a unique character defined by its homogenous (disordered) fabric, or lack of fabric, which creates very thin edges and a brittle character, frustrating many prehistoric technical uses. Within regional sources of obsidian, some are less brittle than others, which is one reason that, as a knapper, I attempt to create a side-notched projectile point from every source that I sample and report.

There are many examples of obsidian source quality differences worldwide; because I know the North American Southwest best, my example is from there. There are five sources or source groups in the Jemez Mountains volcanic field of northern New Mexico that have been produced from the same evolving magma source over about 8.5 Ma (Shackley 2005, 2021; Shackley et al. 2016). All are, in general, excellent media for tool production, and Cerro del Medio (Valles Rhyolite) is volumetrically the largest obsidian source in the region and occurs prehistorically throughout much of western North America. However, as a media for tool production, it is, in my opinion, slightly inferior (more brittle and with more spherulites) to El Rechuelos Rhyolite obsidian, which is older and seems to have been preferred during the Southwest Archaic, when projectile point production may have been more crucial than during the later agricultural period. So, the selection of obsidian as a raw material could have been a multi-faceted decision-making experience in the past; this point is not necessarily evident with provenance studies alone.

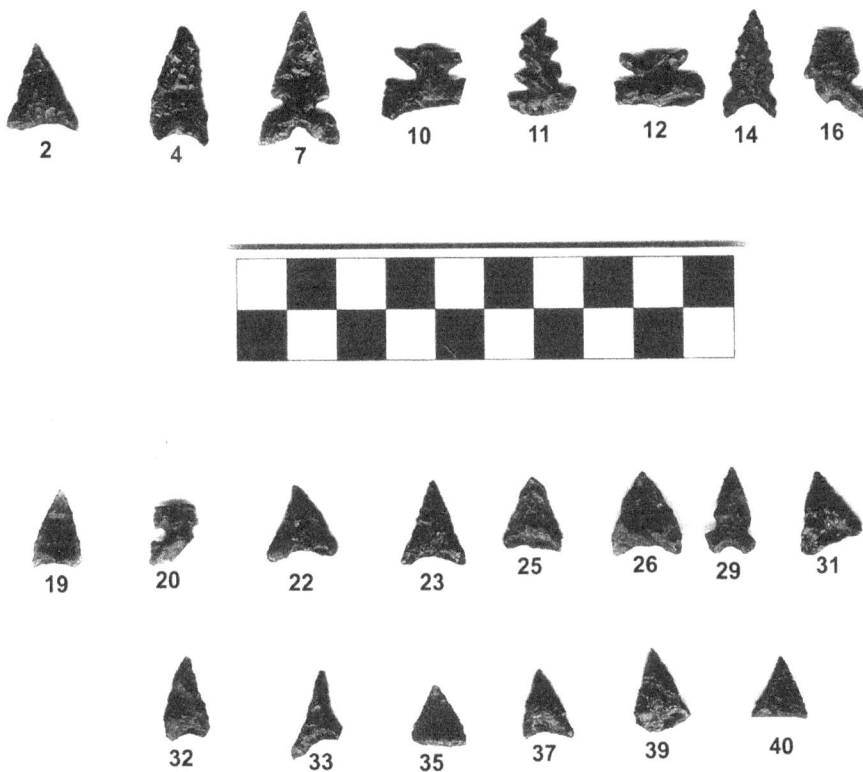

Figure 1. Obsidian projectile points from CA-SDI-9441 (Noble Creek Site) in the eastern Laguna Mountains, San Diego County, California, on the western edge of the Colorado Desert. Numbers 10, 11, and 12 are typical Dos Cabezas Serrated points. Note the breakage pattern, often through the haft element or the deep serrations. Point number 10 is produced from the Tinajas obsidian source in northern Baja California, approximately 150 km south, and numbers 11 and 12 from Obsidian Butte, approximately 85 km east.

A more detailed example comes from the region of southern Alta California and northern Baja California. Here, a unique projectile point type called "Dos Cabezas Serrated", a deeply serrated (barbed), mostly straight to convex-based arrowpoint, is invariably produced from obsidian and most common in the western Colorado Desert in the Peninsular Range granite boulder mountains (McDonald 1992; Shackley 2019a, 2019b; Figure 1). Late Prehistoric Kumeyaay territory is very large (200 km east-west and 250+ km north-south), including much of southern California and northern Baja California, from the desert through the Peninsular Range to the coast, with two major obsidian sources—Obsidian Butte in far southeastern California near the US/Mexico border and Tinajas in northeastern Baja California—as well as a smattering of Western Great Basin sources appearing in all time periods, such as one of the source groups in the Coso Volcanic Field (Hughes and True 1985; Panich et al. 2017; Shackley 2019a). When recovered archaeologically, the obsidian point is often broken through the haft element and only the base recovered (Figure 1), making the potential use for this point a conundrum. In the 1970s, while working in archaeological sites in the Colorado Desert, I had the opportunity to speak with a 100+-year-old Kumeyaay male, Romaldo LaChaapa, and he solved the problem. That boulder-strewn, mountain desert territory is and was inhabited by the crepuscular Desert Bighorn Sheep. Like many artiodactyls, the Desert Bighorn does not sweat to cool, and so Kumeyaay hunters would hunt them during the morning (since it is very difficult terrain to traverse on a hot day), tracking them through blood spatters on the boulders. Romaldo described the projectile point in question perfectly: "an arrowhead, always made from obsidian, with large barbs that would create great damage and much blood." In this case, the technological attributes of the Dos Cabezas Serrated style required obsidian, which was expected to break in the animal. The haft elements recovered archaeologically were a result of planned breakage, and would have still been attached to arrows and apparently discarded. In this particular case, the melding of Indigenous oral history, projectile point technology (prehistoric and modern perception thereof), and specific raw material choice (obsidian) all came together to explain an archaeological conundrum. We often need to look beyond provenance to understand prehistory: many in this volume do just that, and some are beyond my early attempts (see Ward 1977; Shackley 1989).

The provenance and experimental papers in this volume were produced by pXRF data. The "portable XRF revolution" has been a double-edged sword, as has been discussed in print for a while now. The major difference between lab XRF and pXRF, as I see it—besides a more limited capture of the periodic table with pXRF—is spot size. I cannot emphasize enough that to capture variability both within a source and an archaeological sample, as a reflection of melt composition, with pXRF, it would be useful to measure multiple spots on a sample even though it may take a bit longer. This is worth an experiment, say, taking 10 or more samples (shots) from an obsidian artifact (both sides), and looking at the mean and central tendency of the results. (Perhaps someone has done this, and I hope I have not missed that.) As I have noted, most laboratory energy dispersive X-ray fluorescence (EDXRF) instruments (i.e., ThermoScientific Quant'X) irradiate a 25–30 mm area with 8 mm collimation, which provides the same information as multiple shots by a pXRF instrument (c.f. Hughes 2010; and Chapter 8, this volume). I wonder at times if some of the differences between pXRF results and those of source standard data acquired with laboratory XRF are due to single shots of a sample with pXRF (not all perhaps, but some). This, of course, is assuming that comparable instrumental precision is improving with pXRF. I have always been open to running any problem samples, as I have done with some authors in this volume many times—not that it always solves the issues, but it is usually worth a try.

Finally, I would say that obsidian provenance and technological studies are alive and well, even though we as a sub-sub-discipline are small in number. Sharing data, and making it available through open-source instruments, is the way of the future. The scholars in this volume show us the way.

M. Steven Shackley

Geoarchaeological XRF Laboratory, Albuquerque, New Mexico, USA

Department of Anthropology, University of California, Berkeley, USA

References

Hughes, Richard E.

2010   Determining the Geologic Provenance of Tiny Obsidian Flakes in Archaeology Using Nondestructive EDXRF. *American Laboratory* 42: 27–31.

Hughes, Richard E., and Delbert L. True

1985   Perspectives on the Distribution of Obsidians in San Diego County, California. *North American Archaeologist* 6: 325–339.

Le Bourdonnec, F-X, M. Orange, and M. S. Shackley (editors)

2024   *Sourcing Obsidian – A State-of-the-Art in the Framework of Archaeological Research.* Interdisciplinary Contributions to Archaeology. Springer, New York.

McDonald, A. Margaret

1992   *Indian Hill Rockshelter and Aboriginal Cultural Adaptation in Anza-Borrego Desert State Park, Southeastern California.* PhD dissertation, Department of Anthropology, University of California, Riverside.

Panich, Lee, M., M. Steven Shackley, and Antonio Porcayo Michelini

2017   A Reassessment of Archaeological Obsidian from Southern Alta California and Northern Baja California. *California Archaeology* 9: 53–77.

Shackley, M. Steven (editor)

1998   *Archaeological Obsidian Studies: Method and Theory.* Advances in Archaeological and Museum Science 3. Springer/Plenum, New York.

Shackley, M. Steven

1989   *Early Hunter-Gatherer Procurement Ranges in the Southwest: Evidence from Obsidian Geochemistry and Lithic Technology.* PhD dissertation, Department of Anthropology, Arizona State University, Tempe.

2005   *Obsidian: Geology and Archaeology in the North American Southwest.* University of Arizona Press, Tucson.

2011   An Introduction to X-Ray Fluorescence (XRF) Analysis in Archaeology. In *X-Ray Fluorescence Spectrometry (XRF) in Geoarchaeology*, edited by M. S. Shackley, pp. 7–44. Springer, New York.

2019a   Natural and Cultural History of the Obsidian Butte Source, Imperial County, California. *California Archaeology* 11: 21–44.

2019b   The Patayan and Hohokam: A View from Alta and Baja California. *Journal of Arizona Archaeology* 6: 83–98.

2021   Distribution and Sources of Secondary Deposit Archaeological Obsidian in Rio Grande Alluvium New Mexico, USA. *Geoarchaeology* 36: 808–825.

Shackley, M. Steven, Fraser Goff, and Sean G. Dolan

2016   Geologic Origin of the Source of Bearhead Rhyolite (Paliza Canyon) Obsidian, Jemez Mountains, Northern New Mexico. *New Mexico Geology* 38: 52–65.

Ward, Graeme

1977   On the Ease of "Sourcing" Artefacts and the Difficulty of "Knowing" Prehistory. *New Zealand Archaeological Society Newsletter* 20: 188–194.

CHAPTER 1

Introduction (Editor's Preface)

LUCAS R. M. JOHNSON, KYLE P. FREUND, AND NICHOLAS TRIPCEVICH

Lucas R. M. Johnson Far Western Anthropological Research Group, Inc., 1180 Center Point Dr., Suite 100, Henderson, NV 89074, USA

Kyle P. Freund Far Western Anthropological Research Group, Inc., 1180 Center Point Dr., Suite 100, Henderson, NV 89074, USA; Adjunct Assistant Professor at the University of Nevada, Las Vegas, USA

Nicholas Tripcevich Archaeological Research Facility (ARF), University of California, Berkeley, 2251 College Building, Berkeley, CA 94720-1076, USA

Background to this volume

This volume results from the 2021 International Obsidian Conference (IOC), a virtual symposium held in the spring of 2021. Originally scheduled as an in-person event at the University of California, Berkeley, the conference transitioned to a virtual venue due to the ongoing COVID-19 pandemic. We thank all of the more than 70 participants who stuck with us to make the conference a success. Indeed, a wide range of time zones were represented, with many dedicated participants staying up into the wee hours of the night to hear the presentations and engage with others who share a passion for obsidian.

Obsidian studies are a robust facet of archaeology, and this volume highlights a diverse range of research themes. There are seven chapters featuring studies from across the globe (Figure 1), which we have organized broadly by region, including Europe, Africa, Central America, and South America. A separate section with an additional two chapters presents methodological developments in the field.

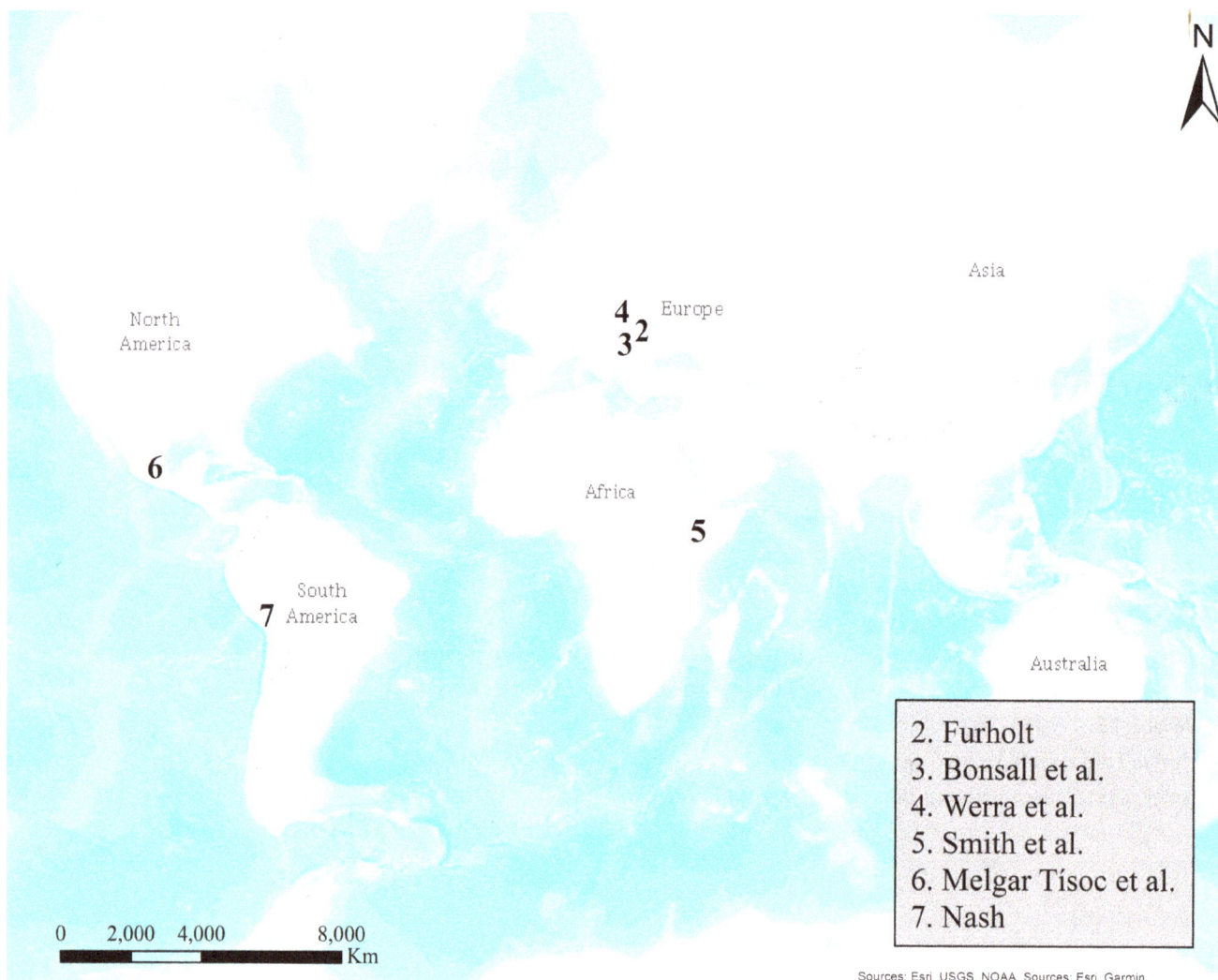

Figure 1. Map displaying the research locations of chapters in this volume.

Europe

The first three chapters of this volume come from Europe, including two from the Carpathian Basin and one from modern-day Poland. In Chapter 2, Furholt provides a broad review of obsidian use in the Neolithic, discussing its varied roles in settlement contexts, burials, and ritual hoards in the Carpathian Basin. The author focuses on Neolithic concepts of value and the economic, social, and ritual functions that obsidian played in society. Such an approach is underdeveloped in the region, and this study advances new understandings of how obsidian was integrated into the lives of the people who used it.

In Chapter 3, Bonsall and colleagues use X-ray fluorescence (XRF) spectrometry to determine the sources of raw materials from the tell site of Hódmezővásárhely-Gorzsa in southeastern Hungry. The site is situated on a large floodplain and is devoid of ready tool stone; thus, they assert that nearly all chipped stone materials come from a minimum distance of 60 km or more. The authors employ two XRF analyzers on two sets of artifacts and explain the difficulties in analyzing artifacts with surface contaminants. Their analyses detail prior hypotheses concerning obsidian source use in the region and outline the relative proportions of certain Carpathian sources over others.

In Chapter 4, Werra and colleagues provide insights into obsidian procurement and technology from a large collection of Neolithic material from the Opatów site in Southeast Poland. The authors examine artifacts from excavations conducted over a period of 50 years and demonstrate that important contributions can be made by bringing new methods and current analytical and theoretical perspectives to existing collections. The technological analysis shows that production was focused on creating blades and flakes with little evidence of formal tool production, and a laborious effort at refitting succeeded in reconstructing lithic reduction sequences. Geochemical analysis using XRF spectrometry concluded that all of the obsidian was from the Carpathian 1 source, located to the south of the site in Slovakia.

Africa

In Chapter 5, Smith and colleagues chemically characterize a major obsidian source to contextualize obsidian use from a ca. 50,000-year-old site in southwestern Ethiopia that corresponds to an era when modern humans were migrating out of Africa. The rockshelter site, Mochena Borago, is located in the southwestern highlands of Ethiopia, a likely refugia during periods of increased aridity. The authors compile XRF data from more than 10 years of research, pointing out the increased precision of modern instrumentation when it is compared to early portable XRF (pXRF) and adjusting minor errors in recording spectra due to different operators. They compare the results with a recent study by Shackley and Sahle (2017) to discuss obsidian in the southern Rift and its use by hunter-gatherers practicing various forms of mobility and technological organization through time.

Central America

In Chapter 6, Melgar Tísoc and colleagues build upon previous neutron activation analysis (NAA) studies on obsidian from Tenochtitlan, specifically from the ritual offering at the Templo Mayor. Using pXRF, they focus on obsidian jewelry and other finely made lapidary items. Applying various statistical methods to differentiate obsidian sources, the authors discuss the significance of material from distant sources brought to the ritual center of Tenochtitlan during the Postclassic period. The authors specifically use new sourcing data to argue that at least one object associated with the Coyotlatelco occupation (such as Tula or Azcapotzalco) represents a relic gifted to ritual practitioners at Templo Mayor.

South America

In Chapter 7, Nash discusses the distribution of a distinctive bifacial artifact known as the "Classic Wari Laurel Leaf Point" within the Wari Empire of South America. As the earliest empire in South America, Wari and their initial expansion to cover over 1,000 km along the Andes and the adjacent Pacific coast is a subject of great interest, one that has largely been documented through the extent of distinctive Wari architectural and ceramic material culture. Nash presents the view that these large foliate points are similarly representative of Wari but have additional characteristics that make their circulation of even greater utility to archaeologists examining the expansion of the original state-level polity in the region. First, as obsidian is geochemically sourceable, it is known that the Quispisisa-type material that dominates Wari assemblages was also circulated widely among Wari non-elite sites far from the geological source. Second, the preforms of these foliate bifaces served as large cores, and the flakes struck from these cores enter into the material sphere of the Wari commoner populace to a much greater degree than decorated pottery.

Nash goes on to present evidence from sites in Moquegua, a far southern outpost of Wari, where she has worked for many years on two adjacent sites of Cerro Baúl and Cerro Mejia. She finds potential evidence of a patron-client relationship between Wari elites and subordinates in the form of smaller obsidian points or exhausted cores being circulated among clients and eventually discarded in non-elite contexts.

Methodological developments in the discipline

Chapters 8 and 9 outline two innovative studies that push the boundaries of how we process geochemical data and use them to address questions of archaeological and geological significance. In Chapter 8, Johnson and colleagues contend with a persistent problem in XRF analysis: small artifact sourcing. They provide a brief overview of the limitations of XRF before describing a revised method for confidently sourcing small artifacts that do not meet conventional assumptions of infinite thickness. More specifically, the authors demonstrate how 95% confidence regions can be applied within ternary diagrams to encapsulate a fuller range of expected variation in a particular geological source regardless of specimen size, allowing for more confidence in making source assignments on smaller artifacts from a wider range of archaeological contexts.

In Chapter 9, Foresta Martin and colleagues discuss a novel approach to directly dating obsidian, highlighting how the ratio of chlorine to sodium decreases with the age of emplacement of obsidian outcrops. This dating technique is successfully applied to obsidian from the Sierra de Las Navajas (Mexico) source, showing a close alignment between the source's estimated dates and those calculated using the new method. Such developments are clearly intriguing, and we hope to see it applied to other archaeological settings around the globe.

In Chapter 10, Tripcevich and colleagues describe an open data hosting platform for South American obsidian source geochemistry that could serve as a template for other regions. They share the NAA data for sources on the OpenContext publishing service but describe an approach for gathering many XRF runs for each chemical group from similarly calibrated instruments using version control.

Continuing a tradition

The Contributions of the Archaeological Research Facility (ARF) at UC Berkeley continues to be a repository for global scholarship and a venue for often niche topics, such as focusing on a single material (see Hughes 1984 and 1989; Gilreath and Hildebrandt 1997). The Contributions of the ARF volumes and those extending from them include studies that form the basis for how obsidian is studied geochemically and anthropologically in the US, extending those prior seminal works from studies abroad (Cann and Renfrew 1964; Hughes 1986; Shackley 2010; Dillian 2014). When the organizers of the 2021 IOC considered a place to hold and then to publish a portion of the proceedings, the ARF and its open access option was the first mentioned due to its history of advanced original scholarship and focus on obsidian from its very first volume (Heizer et al., 1965). Nearly 60 years later, we continue this legacy.

UC Berkeley was part of the first wave applying geochemical techniques to artifacts, in which ARF-affiliated archaeologists collaborated with nuclear scientists in the use of NAA at the Lawrence Berkeley National Laboratory (LBNL) and of spectrometers under development by UC Berkeley's Department of Earth & Planetary Sciences. As discussed by Boulanger and colleagues (2021) during the 2021 IOC conference, the geochemical research initiated by nuclear chemist Isadore Perlman and his collaborators Fred Stross, Helen Michael, and Frank Asaro at LBNL spanning the 1960s to the early 1990s was part of the initial efforts in archaeometric geochemistry focused on ceramics (Asaro and Adan-Bayewitz 2007). An archive of this work was transferred to the Archaeometry program at the Missouri University Research Reactor (MURR) center and later disseminated by the Digital Archaeological Record (tDAR) repository (Boulanger 2013, 2014, 2017).

Geochemical research continued at UC Berkeley, notably in the obsidian geochemical work by ARF founder Robert F. Heizer and his students who, collaborating further with scientists at LBNL and UC Berkeley, initiated some of the earliest XRF analyses of obsidian from California and Mesoamerica (Heizer et al., 1965). This collaboration between anthropologists and LBNL scientists continued in the research of Richard L. Burger, then a graduate student, who coordinated an ambitious obsidian geochemistry study with Frank Asaro involving over 1,000 samples from the Andes and who succeeded in outlining the major obsidian types and consumption patterns in prehispanic Peru and Bolivia (Burger and Asaro 1977, Burger and Asaro 1978).

UC Berkeley and, by extension, the ARF saw a directed effort to advance obsidian studies in the early 1990s. In 1991, M. Steven Shackley arrived at the Phoebe Hearst Museum at UC Berkeley where he continued his systematic geochemical XRF program focused on the US Southwest and grew it into an analytical geochemical service focusing on obsidian and analyzing thousands of samples. An important contribution of Shackley and his contemporaries (e.g., Craig Skinner, Richard Hughes, Mike Glascock) was the pursuit of empirical calibration between similar XRF instrumentation (primarily, the Thermo Fisher Quant'X benchtop XRF). Long an advocate of analytical transparency, most of Shackley's XRF work is available on the ARF's eScholarship open access repository (https://escholarship.org/uc/item/75c689n2) hosted by the University of California. Shackley retired from UC Berkeley in 2012, but his work carries on at the solar-powered Southwest Geochemistry Laboratory in Albuquerque, New Mexico.

Continuing a tradition of publishing within an institution like the ARF is intentional. It enables students, young and experienced, to follow the course of technological change, advancement, and retrospection. New and expanding research on obsidian necessarily learns from that which came before it, and the editors and contributors of the volume and the 2021 IOC uphold this educational journey.

References

Asaro, F., and D. Adan-Bayewitz
    2007   The History of the Lawrence Berkeley National Laboratory Instrumental Neutron Activation Analysis Programme for Archaeological and Geological Materials. *Archaeometry* 49(2): 201–214.

Boulanger, Matthew T.
    2013   Salvage archaeometry: lessons learned from the Lawrence Berkeley Laboratory archaeometric archives. *The SAA Archaeological Record* 13(1): 14–19.
    2014   Lawrence Berkeley National Laboratory (LBNL) Nuclear Archaeology Program Archives, https://core.tdar.org/collection/7901/lawrence-berkeley-national-laboratory-lbnl-nuclear-archaeology-program-archives, accessed January 24, 2023.
    2017   Recycling data: Working with published and unpublished ceramic-compositional data. In *The Oxford Handbook of Archaeological Ceramic Analysis*, edited by A. Hunt, pp. 73–84. Oxford University Press, Oxford.

Boulanger, Matthew T., Nicholas Tripcevich, Richard L. Burger
    2021   Digitization and Preservation of Legacy Datasets: Continued Adventures in Salvage Archaeometry. Paper presented at the 2021 International Obsidian Conference. Berkeley, California.

Burger, Richard L., and Frank Asaro
    1977   *Trace Element Analysis of Obsidian Artifacts from the Andes: New Perspectives on Pre-Hispanic Economic Interaction in Peru and Bolivia.* Technical Information Department, Lawrence Berkeley Laboratory. University of California, Berkeley.
    1978   Obsidian Distribution and Provenience in the Central Highlands and Coast of Peru during the Preceramic Period." In *Studies in Ancient Mesoamerica, III*, edited by John A. Graham, 61–83. Contributions of the University of California Archaeological Research Facility #36. University of California, Berkeley.

Cann, Johnson Robin, and Colin Renfrew
    1964   The Characterization of Obsidian and Its Application to the Mediterranean Region. In *Proceedings of the Prehistoric Society*, Vol. 30, edited by J. G. D. Clark, pp. 111–133. Cambridge University Press, Cambridge.

Dillian, Carolyn D. (editor)
    2015   *Twenty-Five Years on the Cutting Edge of Obsidian Studies: Selected Readings from the IAOS Bulletin.* International Association for Obsidian Studies, Bulletin 52.

Gilreath, Amy J., and William R. Hildebrandt
    1997   *Prehistoric Use of the Coso Volcanic Field.* Contributions of the University of California Archaeological Research Facility #56. University of California, Berkeley.

Heizer, Robert F. (editor)

1965   *Sources of Stones Used in Prehistoric Mesoamerican Sites.* Contributions of the University of California Archaeological Research Facility #1. University of California, Berkeley.

Hughes, Richard E.

1984   *Obsidian Studies in the Great Basin.* Contributions of the University of California Archaeological Research Facility #45. University of California, Berkeley.

1987   *Diachronic Variability in Obsidian Procurement Patterns in Northeastern California and Southcentral Oregon.* Publications in Anthropology, Vol. 17. University of California Press, Berkeley.

1989   *Current Directions in California Obsidian Studies.* Contributions of the University of California Archaeological Research Facility #48. University of California, Berkeley.

Shackley, M. Steven (editor)

2010   *X-Ray Fluorescence Spectrometry (XRF) in Geoarchaeology.* Springer Science & Business Media, New York.

CHAPTER 2

Depositional Patterning of Obsidian Artifacts: Studying Diverse Value Concepts in the Neolithic Carpathian Basin

KATA FURHOLT

Kata Furholt Kiel University, Kiel, Germany (kata.furholt@ufg.uni-kiel.de)

Abstract

Stone tools, although one of the most abundant facets of the archaeological record, have in the past almost exclusively been considered with regards to the transmission of technological traditions or cultural habits, expressed in the presence of "cultures" or "technocomplexes," and too infrequently studied for their role within economic systems and systems of value. In the European Neolithic research tradition, studies of social organization usually focus on exotic materials such as obsidian, jade, lapis lazuli, spondylus shells, early copper, or elaborate pottery. "Exotic," in this context, means materials that have a recognizable visual appearance or that occur rarely, and with original sources that are well-known and can be clearly delineated (e.g., one mountain, one mine) or restricted to a small area. In this sense, obsidian provides an excellent opportunity to look closer at the provenance approach to identifying the potential value of material in the past. To systematically examine the quantitative distribution and exchange of obsidian tools and their integration into community-specific systems of value is an approach that will help promote a better understanding of obsidian's social and economic role in prehistory. For this reason, this chapter focuses on the appearance of obsidian artifacts in a number of different archaeological contexts, including settlement features, burials, and deposits (depots or hoards), to study the various forms of value in the Carpathian Basin.

Introduction

The primary focus of lithic studies lies on provenance, typological and technological analysis, and their temporal and spatial distribution. During the last few decades, the quantitative and qualitative lithic data of Europe has significantly increased, enabling research into lithic materials and the technologies connected to them to lend significance to wider social questions (Scharl et al. 2021: 115, 120–124). This chapter uses the provenance data of lithics in the Carpathian Basin—especially obsidian—to focus on a number of specific activities related to the lithic artifacts from a social archaeological perspective.

Activities such as procurement, production, use, exchange, and deposition are related to the economic, social, and ritual realms of community life. Traditionally, the economic aspect of lithics has been predominant in research on regional or cross-regional exchange systems and direct or indirect networks between communities. In contrast, this chapter pays more attention to the social and ritual activities of early farmers and the largely egalitarian communities of the Neolithic period in the Carpathian Basin. I compare the first farmers (Starčevo-Körös-Criş), agrarian Linearbandkeramik/Linear Pottery, i.e., LBK (Transdanubian LBK, Alföld LBK, Bükk, Szakálhát, etc.), post-LBK communities (Lengyel), and tell-builder societies (Tisza, Vinča). All of these communities had different raw material strategies correlated with their surrounding landscapes, and I present several case studies from different archaeological contexts to study the archaeological evidence of domestic and ritual activities involving lithic materials. For purposes of this chapter, the Hungarian chronological terminology is used, where the LBK period is dated to the Middle Neolithic (while it is Early Neolithic in the Slovakian chronology). Regarding the spatial and temporal scale of the raw material distribution, absolute dates or millennia BC are adopted.

Short overview of Carpathian obsidians

Obsidian is one of the most prominent lithic raw materials not only in the Carpathian Basin but also in Neolithic Southeast and Central Europe. Its continuous use can be traced from the Middle Paleolithic to the Copper Age (Biró 2014: 54). The importance of this raw material was already noted in 1876 at the International Congress of Anthropology and Prehistoric Archaeology held in Budapest and was first studied by Flóris Rómer, József Szabó, and Gyula Szádeczky-Kardoss in the late nineteenth century (Biró 1981: 194; 2004: 4). However, the systematic study of obsidian only began in the mid-1960s, when the obsidian varieties of the western Mediterranean and the Aegean were distinguished and classified, a field of research that has remained continuous since the 1970s (Biró 1981: 194–195; 1984: 47; 2004: 4; Kasztovszky and Biró 2004: 4). Otto Williams Thorpe first undertook the identification and localization of the obsidian varieties of the Carpathian Basin during his fieldwork conducted for archaeometric purposes in 1974–1975. He distinguished two main types: Carpathian 1 (Slovakian obsidian) and Carpathian 2 (Hungarian obsidian) (Williams and Nandris 1977; Williams Thorpe et al. 1984). This research was continued by Katalin T. Biró from the 1980s, as well as by György Szakmány and Zsolt

Kasztovszky who submitted samples for various archaeometric analyses (Kasztovszky and Přichystal 2018). Biró distinguished two main varieties among the Hungarian obsidians. She retained Williams Thorpe's Carpathian 1 and 2 groups (abbreviated as C1 and C2) but added more details to these two major types.

The most significant quantities of obsidian can be found in the southerly regions of the Tokaj-Eperjes Mountains (Prešovsko-Tokajské Pohoří), which is related to Neogene (Miocene) volcanism (Szepesi et al. 2018: 167–170). However, the best-quality obsidian occurs in the Slovakian section of the mountain range. The C2 type geological sources lie in the Mád-Erdőbénye-Olaszliszka area (C2E type) and in the Tolcsva area (C2T type) (Figure 1), as their potential primary sources are related to the (geological) deposits of Badenian-Sarmatian-Lower Pannonian volcanism (12.8±0.5 and 10.6±0.5 Ma) (Szepesi et al. 2018: 167; Pécskay et al. 1987: 245–251; 2006: 517; Pécskay and Molnár 2002: 310–311; Lexa et al. 2010; Zelenka et al. 2012). Good-quality Slovakian obsidian is repre-sented by C1 and is found in the Viničky (Szőlőske), Mala Bara (Kisbár), and Streda nad Bodrogom (Bodrogszerdahely) areas, which is connected with acid rhyolitic volcanism of Upper Badenian-Lower Sarmatian (15±2 and 12±0.5 Ma) (Kaminská 2018: 101; Bačo et al. 2017: 207–212, 2018: 159–163). Farther to the north, an extraction site exploited

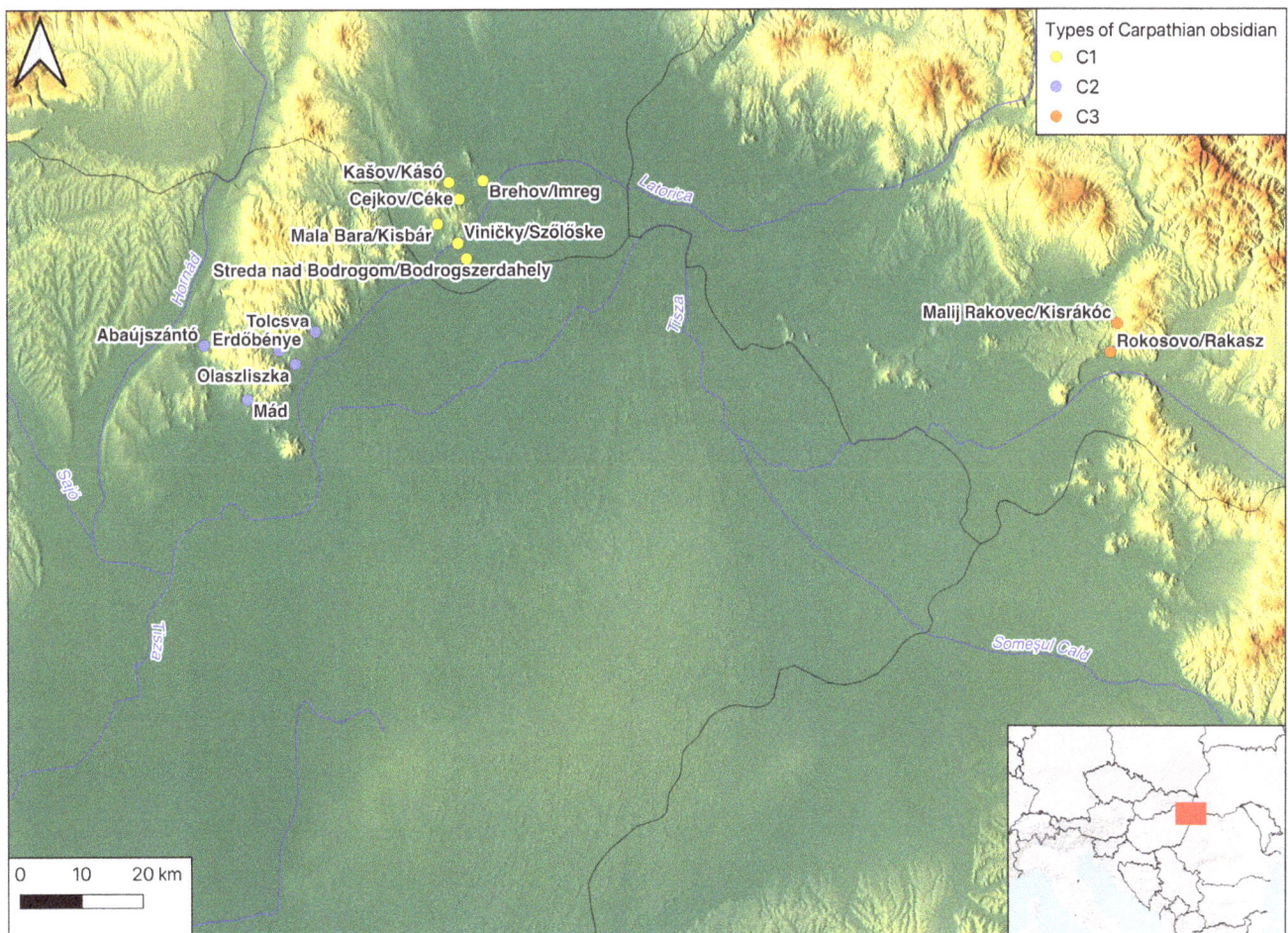

Figure 1. Map of the geological sources of the Carpathian obsidian types.

during the Upper Palaeolithic and the Neolithic is assumed in the Kašov (Kásó), Cejkov (Céke), and Brehov (Imreg) area (Bačo et al. 2018; Bánesz 1991, 1993; Kaminská 2013, 2018; Nandris 1975; Přichystal and Škrdla 2014; Szepesi et al. 2018). Béla Rácz has identified and described a third Carpathian obsidian variety from the Transcarpathian region (Ukrainian Carpathians and Romania), designated as Carpathian 3 (C3) in the literature. The primary geological sources of C3 obsidian can be found in the Rokosovo (Rakasz) and Malij Rakovec (Kisrákóc) areas in Transcarpathian Ukraine (Rácz 2008: 51–52; 2009: 322–325; 2012: 352–353, 360; 2013: 132–133; 2018; Rhyzov 2018).

These sources supplied the Carpathian Basin and its neighboring regions with obsidian (Biró 1981: 194, 203). The Hungarian obsidians are much smaller than the Slovakian ones; they are generally grey or yellowish-grey and covered with a fissured cortex (Figure 2). The larger Slovakian obsidians are covered with a thin, coarse cortex whose color matches the dark grey color of the rock. Slovakian obsidians are transparent or translucent, while Hungarian ones are opaque with a silky luster, a striated patterning, and greyish or black hues (Biró 1981: 201, 203; 1998: 33). Transcarpathian obsidians have a black or greyish, weathered, pitted cortex. The size of the surface obsidian blocks varies; their diameter can be as much as several decimeters. Black pieces with a pitch-

Figure 2. Varieties of Carpathian obsidians.

like luster are particularly frequent, although pieces with greyish hues also occur (Rácz 2009: 324; 2012: 359; 2013: 132–133). The major and trace element composition of the Carpathian obsidians differ: Slovakian obsidian has a higher silicon dioxide and a lower iron content than the Hungarian varieties (Biró 1981: 201; 1998: 33; 2004: 5; Burgert et al. 2017; Kasztovszky and Biró 2004: 5–6; Kohút et al. 2021).

Contextual analysis and research questions

Of all the lithic raw materials, obsidian has the longest and most widely distributed and disseminated research history in the Carpathian Basin. This fortunate situation is supported by a clearly delineated geological source, which has led to exceptional attention by archaeologists who have been able to easily recognize the connection between communities and regions via this raw material. On the other hand, the glassy appearance makes obsidian aesthetically appealing, and the glass-like physical character provides for high-quality knapping. Overall, the well-recognizable visual appearance caught human attention both in the past and present. For this reason, obsidian was a prominent material in the Neolithic exchange network system. Today, obsidian is again drawing a particular research interest, and the vast majority of lithic analyses in the region report and publish on obsidian.

The diachronic changes of obsidian distribution in the Carpathian Basin are well published and backed by geochemical analyses. These publications and datasets provide a rich opportunity to carry out: 1) a comparative analysis to study how obsidian is related to other materials in the exchange networks; and 2) a contextual analysis and close examination of the archaeological context and depositional patterns of obsidian. Thus, in this chapter, three main research questions will be explored:

1. Do depositional patterns of obsidian vary in relation to distance from the source?
2. Can we see differences between domestic, burial, and ritual use?
3. What do patterns of deposition in burials and depots and discard in settlement contexts tell us about different forms of the value of a raw material/artifact?

In the following sections, I will discuss the patterns of obsidian procurement, use, and deposition in the domestic, burial, and ritual context.

Domestic context: obsidian from settlement features

In this section, several case studies are discussed concerning lithic assemblages from LBK and post-LBK/Lengyel settlements to compare regions that are close and far from obsidian raw material sources (Tokaj-Eperjes Mountain).

The Karancsság-Alsó rétek site is associated with LBK (Notenkopf/Music notes, Bükk) (Bácsmegi 2003, 2014a, 2014b) and Lengyel communities, while Szécsény-Ültetés (Soós 1982; Fábián 2005, 2010, 2012; Fábián et al. 2016) and Vráble-Veľké Lehemby

(Furholt et al. 2014, 2020; Müller-Scheeßel et al. 2016) are only related to LBK (Zseliz). The raw material distribution in these sites shows that obsidian and the locally available limnic silicite were the dominant lithic types, and besides them, the Bakony radiolarite (Transdanubian Mid-Mountain) and erratic flint variations appeared in low numbers (Figure 3).

The entire knapping process of obsidian is represented at all three sites, from decortication flakes to unretouched flakes and blades with small-size cores. Retouched tools are more or less missing from the assemblages because the unretouched obsidian blades were probably perfectly functional with their sharp cutting edges. More than 50% of the lithic material in Karancsság and Szécsény is obsidian (Biró 1987: 154–159; 1998: 45; Szilágyi 2009), while in Vráble the obsidian is represented at almost 30% (Cheben et al. 2020: Table 1). The former two sites are located (130–145 km) closer to the obsidian sources than Vráble (235 km).

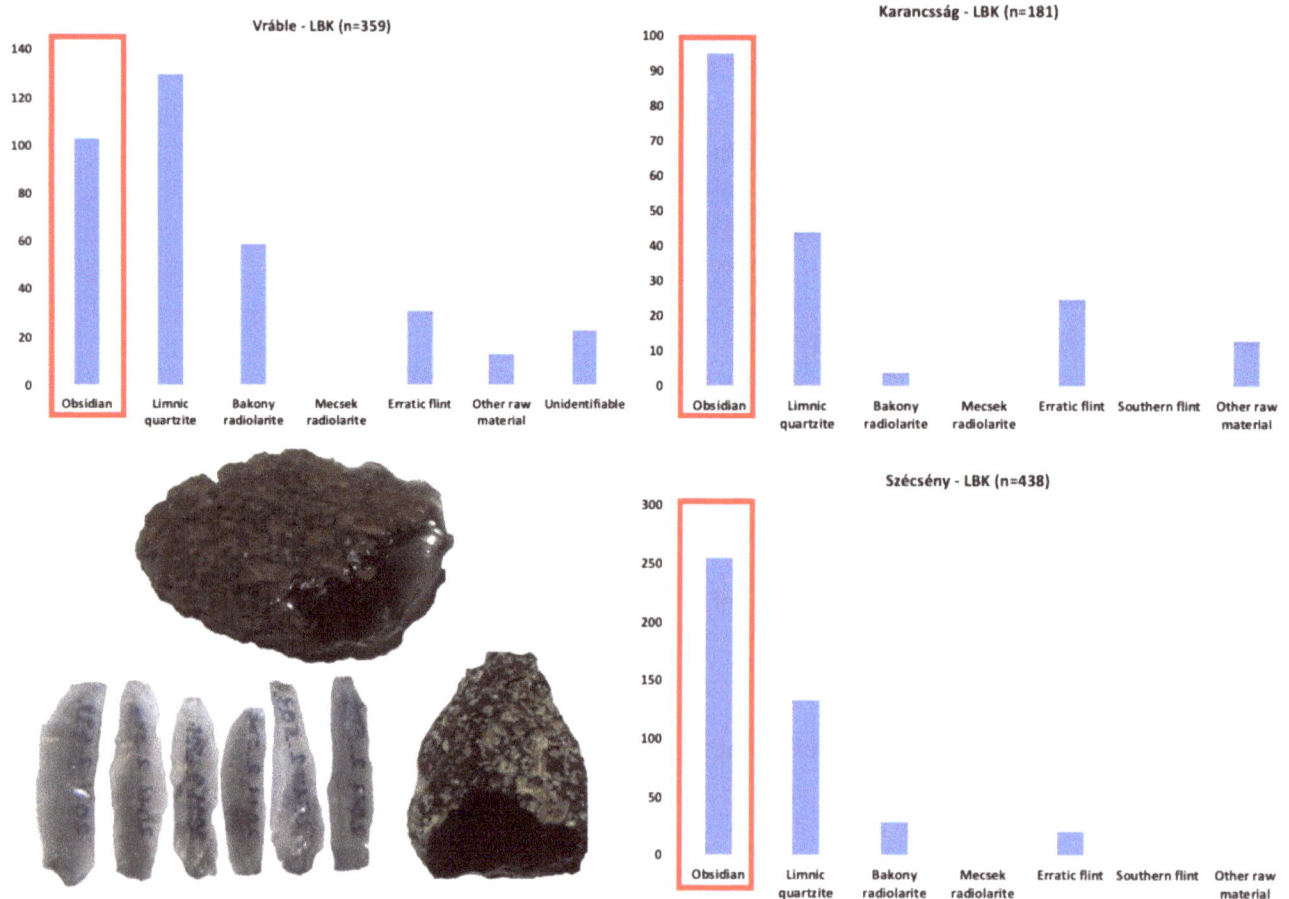

Figure 3. The raw material spectrum of the compared settlement lithic assemblages (Vráble, Karancsság, Szécsény) in the second half of the 6th Millennium BC on the north part of the Carpathian Basin; typical decortication flakes, and unretouched blades from the Karancsság site [bottom left].

Table 1. Raw material distribution of the compared LBK assemblages (Vráble, Karancsság, Szécsény). Data source: Cheben et al. 2020; Szilágyi 2009; Biró 1987: 154–159; 1998, 45.

Raw material	Vráble		Karancsság		Szécsény	
	n	%	n	%	n	%
Obsidian	103	28.69	95	52.49	255	58.22
Limnic quartzite	130	36.21	44	24.31	133	30.37
Bakony radiolarite	59	16.43	4	2.21	29	6.62
Mecsek radiolarite	0	0.00	0	0.00	0	0.00
Erratic flint	31	8.64	25	13.81	21	4.79
Southern flint	0	0.00	0	0.00	0	0.00
Other raw material	13	3.62	13	7.18	0	0.00
Unidentifiable	23	6.41	0	0.00	0	0.00
Total	**359**	**100.00**	**181**	**100.00**	**438**	**100.00**

In the case of the Karancsság and Szécsény assemblages, Neolithic knappers followed LBK knapping traditions and used the blade debitage concept and applied it on obsidian as they would on flints or other materials. The obsidian had better knapping properties than other lithics, which provided a perfect opportunity to prepare only unretouched blades and use them as a tool without further modifications. At the same time, the occurrence of end and edge retouch was common on the less knappable limnic silicites and flints. Thus, the raw material quality had an impact on the technological behavior and helped transform knapping traditions of LBK communities.

The Karancsság lithic assemblage allows for a comparison of LBK and Lengyel raw material use and technological behavior. In the comparison, I integrated the well-published Southeastern-Transdanubian Lengyel lithic assemblages from Alsónyék-Bátaszék, Zengővárkony-Igaz-dűlő, Pécsvárad-Aranyhegy, Villánykövesd, Lengyel-Sánc, and Mórágy-Tűzkődomb (Biró 1989, 1990, 1998) (Table 2). Figure 4 shows that the obsidian is represented in a very low number in the Transdanubian assemblages, located further away from the sources, while it is at almost 60% in Karancsság (Szilágyi 2009), located closer to the sources. The distance from the original source also affects the size of the cores. In Transdanubia, the locally available Mecsek radiolarite dominated the lithic assemblages, and Mecsek radiolarite cores are on average larger than the obsidian, which are mostly represented by finely prepared blade cores or micro-blade cores.

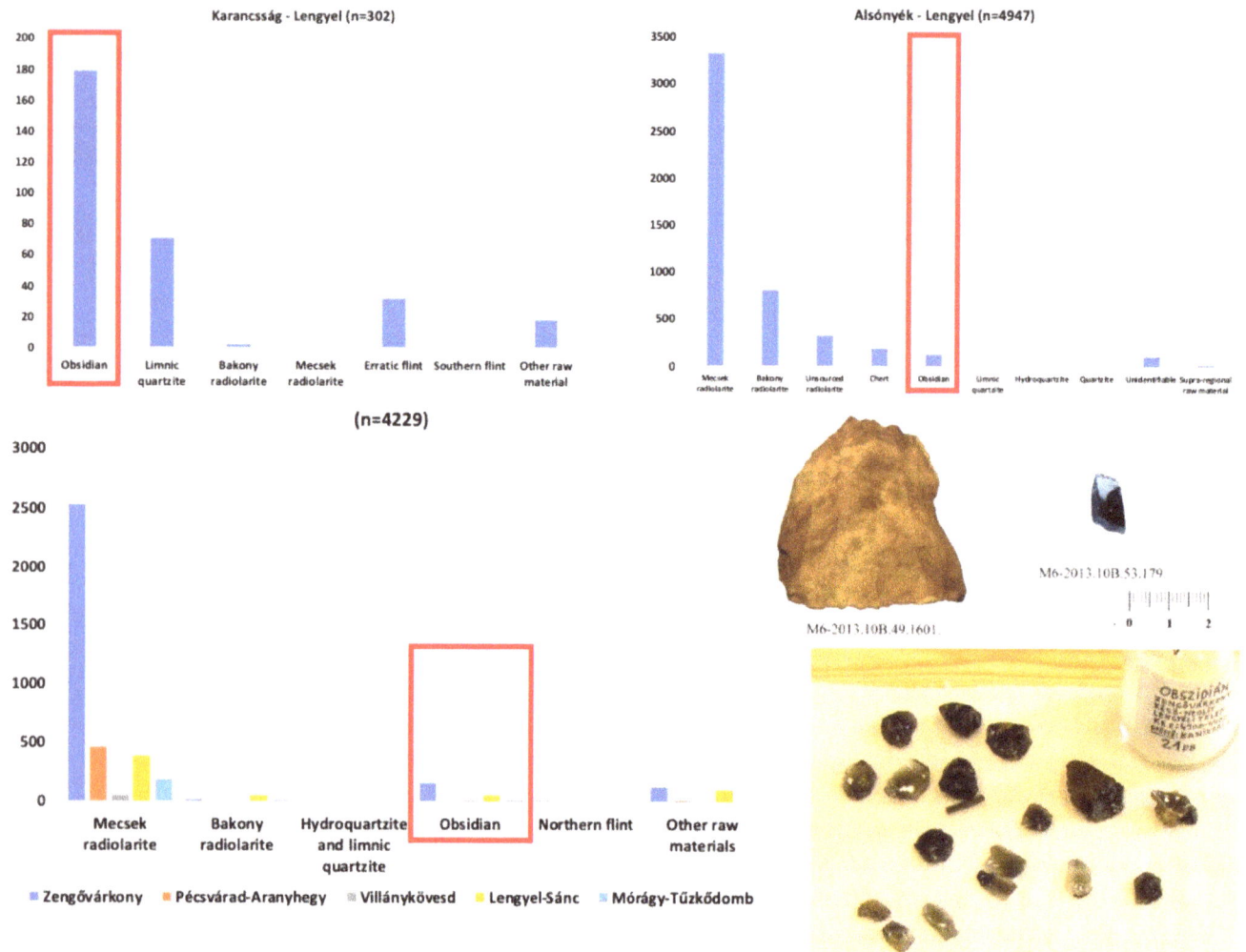

Figure 4. The raw material spectrum of some lithic settlement assemblages in the first half of the 5th Millennium BC (Karancsság [top left], Alsónyék [top right], and Zengővárkony, Pécsvárad, Villánykövesd, Lengyel, and Mórágy [bottom]); an obsidian microblade core from Alsónyék, and some small-size flakes and blades from Zengővárkony.

Thus, besides the local limnic quartzite/silicite or radiolarite, obsidian was quantitatively important in the Northern Hungarian Mid-Mountains area; yet at a greater distance across the Danube in the south Transdanubian region, obsidian was only represented by a few pieces. Even though these communities are likely to have had social connections with other obsidian-rich communities, the raw material selection strategy was different. It seems that the obsidian did not have specific value content. If it had, then we would expect obsidian quantity to have been higher. But it seems it was easily replaceable with other raw materials. Both LBK and Lengyel communities located close to the sources used obsidian in significant quantities, but those farther away did not invest much energy to access obsidian.

Table 2. Raw material distribution of the compared Lengyel assemblages (Karancsság, Zengővárkony-Igaz-dűlő, Pécsvárad-Aranyhegy, Villánykövesd, Lengyel-Sánc, Mórágy-Tűzkődomb). Data source: Szilágyi 2009, 2019a, 2019b; Biró 1989, 1990, 1998.

	Karancsság		Zengővárkony-Igaz-dűlő		Pécsvárad-Aranyhegy		Villánykövesd		Lengyel-Sánc		Mórágy-Tűzkődomb	
Raw material	n	%	n	%	n	%	n	%	n	%	n	%
Obsidian	179	59.27	158	5.56	2	0.41	13	16.67	55	9.17	13	5.78
Limnic quartzite	71	23.51	3	0.11	1	0.21	1	1.28	8	1.33	0	0.00
Bakony radiolarite	2	0.66	20	0.70	6	1.24	4	5.13	49	8.17	13	5.78
Mecsek radiolarite	0	0.00	2530	88.99	462	95.65	53	67.95	386	64.33	191	84.89
Erratic flint	32	10.60	8	0.28	0	0.00	4	5.13	1	0.17	0	0.00
Southern flint	0	0.00	0	0.00	0	0.00	0	0.00	0	0.00	0	0.00
Other raw material	18	5.96	124	4.36	12	2.48	3	3.85	101	16.83	8	3.56
Unidentifiable	0	0.00	0	0.00	0	0.00	0	0.00	0	0.00	0	0.00
Total	**302**	**100.00**	**2843**	**100.00**	**483**	**100.00**	**78**	**100.00**	**600**	**100.00**	**225**	**100.00**

Burial context: obsidian from grave deposits

Lithics as regularly deposited grave goods first appeared in the LBK context in the second half of the 6th millennium BC and became a more or less frequent element of the burial ritual of Lengyel communities in the first half of the 5th millennium BC. Lithic grave goods represent varied forms in both raw material and tool type (from flakes made from local material to Volhynian blades), but flint blades are the most common. The first half of the 5th millennium BC has been interpreted as the period in which we see the first appearance of social inequality, and grave goods are seen as markers of prestige based on their exotic raw material (Siklósi 2013; Zalai-Gaál 1986; 2004; 2005; 2010).

With this in mind, it is remarkable that the supposedly valuable obsidian did appear frequently in burials. Known examples include two huge obsidian blade cores of the second half of the 6th millennium BC, published from Polgár-Ferenci-hát, Grave 867 (Whittle et al. 2013: 75), and one obsidian blade from Vráble, Grave G10/S21 (Müller-Scheeßel and Hukeľová 2020: 182–183, Figure 3.2.16, 226, Pl.3.2.2.) (see Figure 5).

Our knowledge about lithic grave goods and their role in burial rituals in the first half of the 5th millennium BC has significantly increased in recent years, especially through the study of the material at Alsónyék-Bátaszék. Local and regional flint blades or other blade-based tools (e.g., trapeze, end-scraper on blade) were usually placed close to the

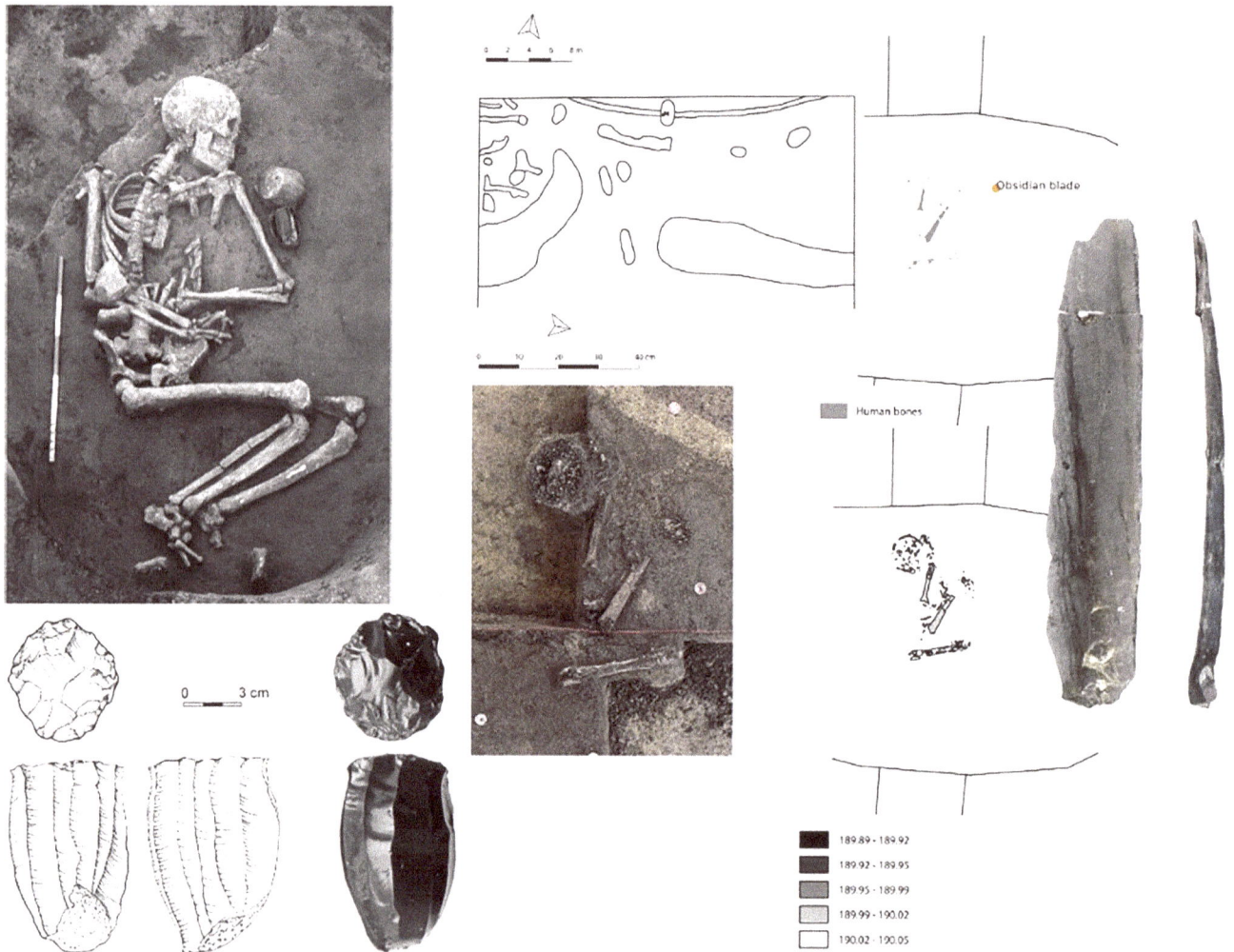

Figure 5. Obsidian tools as grave goods in the LBK burial context in the second half of the 6th Millennium BC. On the left side: Polgár-Ferenci-hát, Grave 867 (Whittle et al. 2013: 75). On the right side: Vráble, Grave G10/S21 (Müller-Scheeßel and Hukeľová 2020: 182–183, Figure 3.2.16, 226, Pl.3.2.2)

deceased in Lengyel communities. The Alsónyék-Bátaszék site displays an outstanding opportunity for studying the role of lithic grave goods, in a statistically representative way, with 2,236 burials, where 33% of the graves had lithic items (Szilágyi 2019a: 184–185; 2019b: 88–89). The raw material spectrum shows a solid local and regional orientation, while the distant raw materials make up a small portion (in the Alsónyék-Bátaszék settlement, 22 pcs., 0.44%, and in the burial assemblages, 60 pcs., 4.57%) (Szilágyi 2019a: 109–110, 120, 293; 2019b).

The chipped stone discovered from the burials at Alsónyék-Bátaszék differs from the settlement assemblages in several ways. The dominance of blades in burials is highly conspicuous, and the tools make up a more sizeable portion of the burial assemblage than in the settlement assemblages. The other lithic categories are represented in negligible proportions, indicating that they were not considered among the artifact types selected for deposition in burials. Blades and tools were made from various raw materials; it is also apparent that being import commodities, the chipped stone of supra-regional raw materials was mainly deposited in burials, not in settlement contexts. All in all, 36 obsidian artifacts were discovered, including 5 flakes, 25 blades, and 6 cores (Szilágyi 2019a: 293). All cores are smaller-sized blade cores, with lengths ranging between 15 and 29 mm and widths between 12 and 23 mm. The obsidian cores were placed on the left side of the upper body, especially the arm bone, and in one case, right beside the mandible (jaw) (Figure 6). Not only are the grave positioning and blade core type similar, but every core is made from the transparent C1 obsidian, which emphasizes the homogeneity of this particular selection for the burial activity. This definite pattern with the obsidian cores in the burial context suggests that it could have had a specific meaning or code of action regarding where and in what form it had to be placed in the grave.

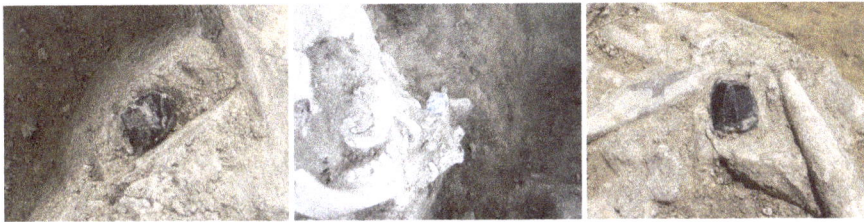

Figure 6. The grave position of obsidian cores in Lengyel burials at Alsónyék-Bátaszék.

Ritual context: obsidian (and non-obsidian) deposits

Obsidian deposits are known from the Upper Paleolithic (Cejkov/Céke) until the Middle Neolithic (Bánesz 1974). One of the most prominent and emblematic finds of the Neolithic period in Hungary—if not the entire Carpathian Basin—is the ensemble of 12 pieces of large-sized conical cores from Nyírlugos. This depot was found by coincidence, which is unfortunately why the Middle Neolithic dating is insecure. Jenő Hillebrand published the Nyírlugos assemblage in 1928 (Hillebrand 1928: 39–42), and it is now presented in the permanent archaeological exhibition of the Hungarian National Museum. The Middle

Neolithic date of the newly published (Biró et al. 2021) four conical cores from Besenyőd is also questionable, as these cores were donated—without original data source—by Lajos Tar (a local teacher) to the Jósa András Museum in 1947 (Biró et al. 2021: 95). Many special obsidian collections—like other kinds of deposits too—often appeared by chance, and unfortunately, we will never know the original archaeological context. We can determine with or without the exact archaeological context that all but three deposits of obsidian objects date to the Middle Neolithic (second half of the 6th millennium BC), and two of those three have an insecure date. In addition, their spatial concentration is clearly restricted to the Upper Tisza Region, which is a relatively small area from the Carpathian Basin perspective (Figure 7).

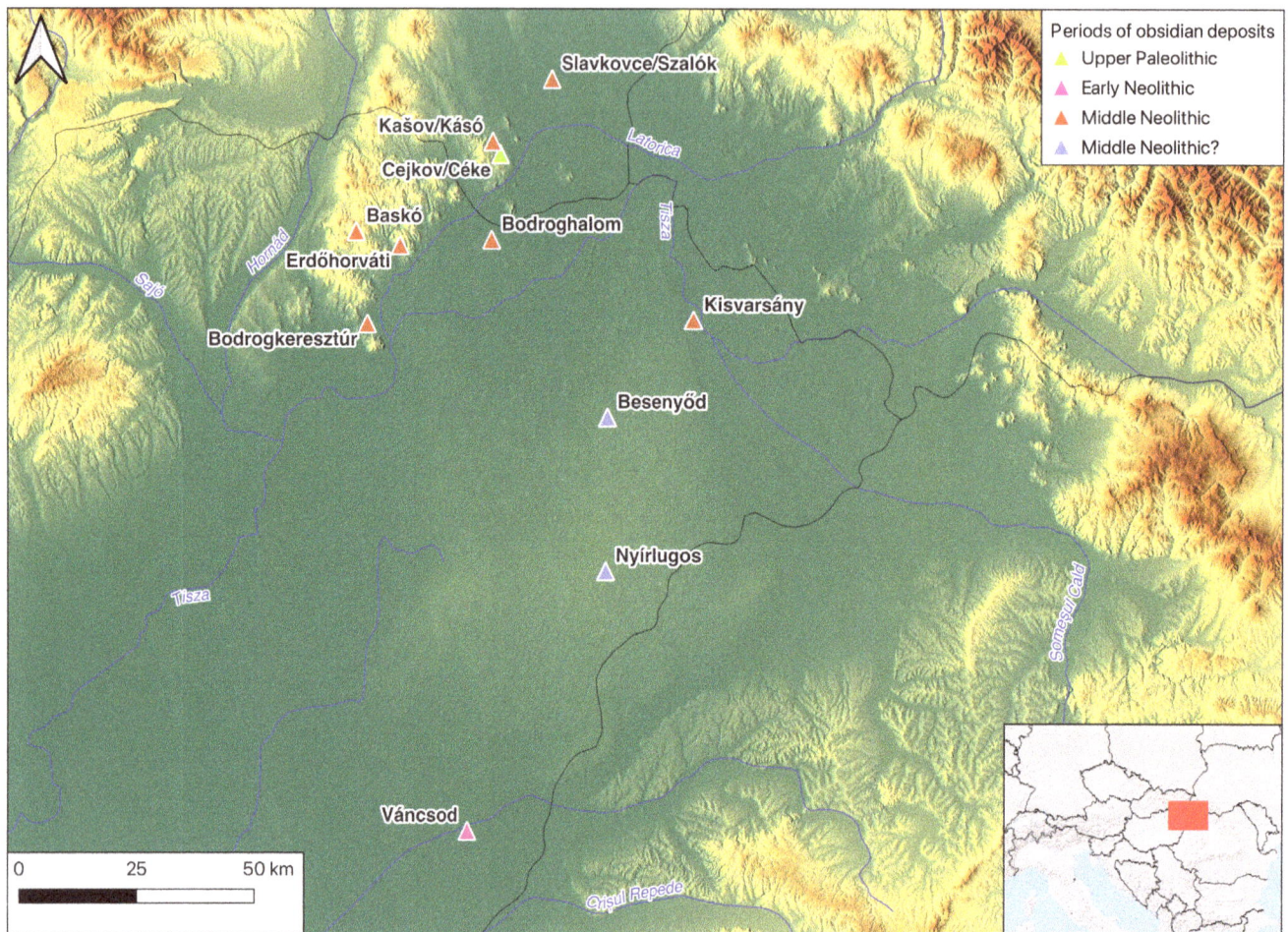

Figure 7. Map of the obsidian deposits in the Upper Tisza Region.

Most of the deposits are located north of the Tisza River, between or near the Hornád (Hernád), Latorica (Latorca), and Bodrog Rivers. Topographically, it is the Tokaj-Eperjes Mountains (also known as Zemplín or Zemplén), located east of the Northern Hungarian Mid-Mountains and southeast of the Western Carpathians, that is the geological source of obsidian. An interesting point concerning this spatial pattern is that it is the previously mentioned, dubiously dated, Middle Neolithic deposits of Besenyőd and Nyírlugos that are situated farther from the original source than most others with a maximum distance of 113 km. This is relevant information for the Váncsod site, where 13 obsidian nodules were discovered at a rescue excavation in 2017 (Priskin et al. 2019: 53). The assemblages are associated with the Early Neolithic Körös population, but at this time, further information about the original archaeological context and other finds has not yet been published. Thankfully, prompt gamma-ray activation analysis (PGAA) was done on the nodules, showing that all of them are made from C1 obsidian, which is approximately 160–170 km away. Váncsod is located in the northern part of the Bihor (Bihar) plain, which corresponds to the Sebes-Körös (Crişul Repede River) alluvial cone. The Váncsod, Besenyőd, and Nyírlugos sites are situated in the Hajdúság and Nyírség, which is a flatland region with sand and meadow soil in the northeast part of the Great Pannonian Plain (Alföld). The rivers have deposited thick sediment in this region along with their alluvial cones and abandoned beds (Kalicz 2011: 45). Besenyőd and Nyírlugos are located 60–70 km away from the Tisza River and running through the area are only a few tiny streams (János, Besenyőd, Lugos, Konyár-Kálló), which may not have existed in this exact location during the Neolithic. The assemblage from Váncsod is 116 km away from the Tisza River, but the Sebes-Körös (Crişul Repede) and the Berettyó Rivers (Barcău) are located just 15 km away.

These three obsidian deposits suggest that the raw material was carried farther from the mountain region and transferred along the smaller rivers and streams. This spatial pattern correlates with the fact that the Neolithic communities in this region also preferred the smaller rivers, which provided an excellent and safe opportunity to live on alluvial ridges. The Tisza River was the main natural transport route for obsidian, spondylus, and copper items in the Neolithic period (Bajnóczi et al. 2013; Siklósi 2013; Siklósi and Csengeri 2011; Siklósi et al. 2017; Kovács 2013). Besides obsidian, objects made from different materials appeared to be significantly more frequent in the Middle and Late Neolithic periods. Thus, the Tisza River and the smaller streams acted as communicational networks between communities.

We now turn to the technological patterns of objects found in obsidian deposits, focusing on homogeneity. So far, the published obsidian deposits show the preparatory phase of the entire knapping activity. In lithic technological research, the characteristics of reduction represent the different production phases, which are shaping, flaking, and retouching (Inizan et al. 1999: 30; Odell 2006: 1–12). We can recognize the tool-making steps and reconstruct the knapping activity based on these technologically well-known traces. The preparatory phase represents core forming for the further removals, which is one of the first steps in the knapping process. This stage of the *chaîne opératoire* is solely represented in the obsidian deposits. Nodules with a more or less untouched natural cortex are considered a selected piece for later knapping activity. Nicely prepared blade cores can also be seen as a preparatory phase of the *chaîne opératoire* because the blade-debitage surface is prepared, but the size of the cores suggests that the entire blade removals did not happen (Allard et al. 2017). In this sense, large blade cores had considerable potential for producing several blades (Figure 8).

The homogeneity of the obsidian deposits can be understood better when compared to other non-obsidian lithic deposits. All non-obsidian lithic assemblages were placed in

Table 3. Overview of the obsidian deposits.

* No distances are given when it is closer than 10 km; the kilometers were counted as approximate straight-line distances as the crow flies. References for Table 1: Priskin, Szeverényi, and Wieszner 2019; Hillebrand 1928; Biró 2014, 1998; Bánesz 1991; Kaczanowska and Kozłowski 1997; Bánesz 1993; Korek 1983; Bella 1920; Nandris 1975.

Site/locality	Subsite name	River*	County
Besenyőd		Tisza (60 km)	Szabolcs-Szatmár-Bereg
Nyírlugos	Erzsébet hegy	Tisza (70 km)	Szabolcs-Szatmár-Bereg
Kisvarsány	Hídér	Tisza	Szabolcs-Szatmár-Bereg
Erdőhorváti	Szelek fej	Bodrog	Borsod-Abaúj-Zemplén
Baskó	Legelő	Bodrog (right beside Tolcsva and Erdőbénye)	Borsod-Abaúj-Zemplén
Váncsod	Szénás dűlő	Berettyó, Körös (16 km), Tisza (115 km)	Hajdú-Bihar
Bodroghalom	Medve tanya	Bodrog	Borsod-Abaúj-Zemplén
Cejkov/Céke		Bodrog (right beside Viničky, Kašov)	Trebišov, Košice
Kašov/Kásó	Čepegov I/Csepegő	Bodrog (right beside Viničky, Cejkov)	Trebišov, Košice
Bodrogkeresztúr	Kutyasor	Bodrog, Tisza (right beside Tokaj)	Borsod-Abaúj-Zemplén
Slavkovce/Szalók		Laborec, Tisza (30 km)	Michalovce, Košice

Country	Age	Culture	Items	Reference
Hungary	MN	Bükk	4 conical cores	Biró-Kasztovszky-Mester 2019
Hungary	MN?	Bükk	12 large conical cores	Hillebrand 1928, Biró 2014, 55, Table 1(a)
Hungary	MN	LBK (Szamos-region painted pottery group)	9 raw material lumps - precores	Korek 1983, Biró 1998
Hungary	MN	LBK (Bükk)	7 nodules-precores	Nandris 1975
Hungary	MN	LBK	unknown	Nandris 1975
Hungary	EN	Körös	14 raw material lumps - precores	Priskin et al. 2019
Hungary	MN	LBK (Bükk)	24 blades	excavatied by M. Hellebrand
Slovakia	UP	Gravettian	41 nodules and precores	Bánesz 1974
Slovakia	MN	LBK (Bükk)	13 large blade cores	Bánesz 1991, 1993
Hungary	MN	LBK (Bükk)	700 flakes and blades	Bella 1920-1922, Biró-Kasztovszky-Mester 2019
Slovakia	MN	Early LBK (Raškovce type)	30 large lumps (altogether 13-14 kg)	Kaczanowska & Kozłowski 1997, 177-253, Biró 1998

Figure 8. Some examples of the obsidian blade core deposits from Kašov-Čepegov [top left and right] (Allard, Klaric, and Hromadová 2017: 22, Figure 4; 24, Figure 6), Besenyőd [bottom left] (Biró, Kasztovszky, and Mester 2021: 96, Figure 1), and Nyírlugos [bottom right] (Kasztovszky, Biró, and Kis 2014: 152, Figure 1).

clay vessels of different styles and forms in the Carpathian Basin. These vessels were for storage; as mostly functional pots, they were not elaborately decorated or not decorated at all. For example, the vessel from the Early Neolithic site Endrőd 39 is heavily worn in the middle and contained 101 pieces of Balkan flint (Biagi and Starnini 2013). The Middle Neolithic site of Boldogkőváralja represents an extreme case with 566 blades fragments made from the locally available limnic silicite (Mester and Tixier 2013; Faragó et al. 2019). In a broken, small-sized jar, 33 pieces of Bakony radiolarite blades were discovered from the Late Neolithic site of Szegvár-Tűzköves (Biró 2009). All non-obsidian lithic deposits are each made from the same raw material and tool type, which can be local or distant knappable raw materials and flakes or blades; thus, the homogeneity of these sets and

Figure 9. Non-obsidian deposits from Endrőd (blue) (Biagi and Starnini 2013: 53, Figure 5), Boldogkőváralja (green) (Mester and Tixier 2013: 176, Figure 5; 178, Figure 6), and Szegvár (red) (Biró 2009) Neolithic sites.

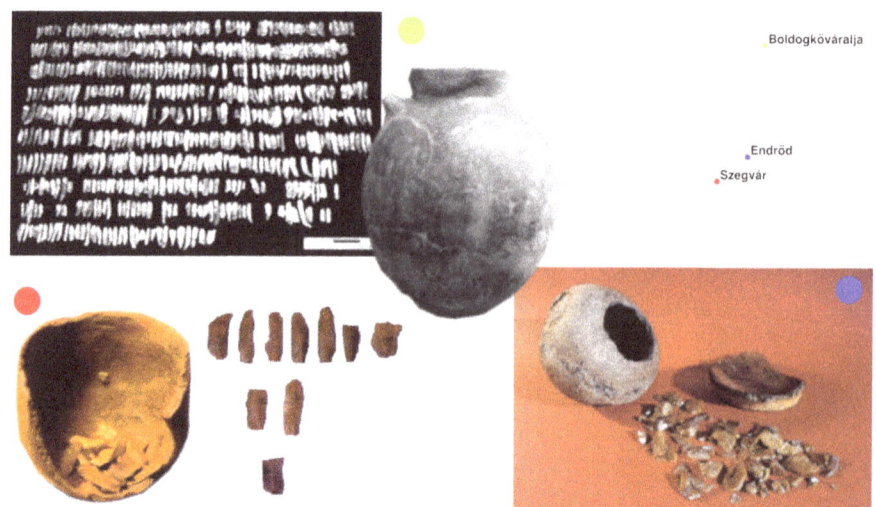

the high number of pieces represent a clear pattern (Figure 9). Such flint blade or mass chipped stone deposition in vessels is also known from the Copper Age context (e.g., Kosd, Pietrele-Măgura Gorgana) (Patay and Marton 2018; Gatsov and Nedelcheva 2015).

Regional patterns of distribution

Many authors emphasize the special significance of the obsidian deposits and refer to the potential explanation of deposits as storage, which more or less relates to the idea of property (some material value hidden away for the "near or further future"). Besides this interpretation, hoarding is another term used to describe this depositional pattern (Biró et al. 2021: 100). The latter bears the additional connotation of a prestigious material, which had a high economic and social value. Both narratives are based on the economic value of obsidian and the suggestion that it was highly valuable for Neolithic communities. We should try to refrain from such ad-hoc interpretations and interrogate the archaeological evidence more closely to recognize and describe the patterns behind the obsidian deposition practices. By doing so, we can hopefully proceed toward a study of the intentions and social contexts behind those acts.

The amount of obsidian found and its spatial distribution show clear patterns in the Neolithic. For this article, I collected data from 219 sites where obsidian appeared in a lithic assemblage. Besides the fact that a database is never complete, I would like to give a non-representative overall ratio of the obsidian distribution during the Neolithic: 14% of the Early Neolithic sites ($n = 30$), 40% of the Middle Neolithic sites ($n = 86$), and 24% of the Late Neolithic sites ($n = 52$) have obsidian in their lithic assemblages. In the database, 11 sites are dated to the Neolithic without any precise time period (5% of the collected sites), 12 sites are not dated (5% of the collected sites), 24 (many of them Middle Neolithic) sites are multi-period, 1 site is Upper Paleolithic, and 3 sites are related to the Copper Age (Table 4).

Table 4. Period information about the collected sites where obsidian appeared.

Periods	Number of sites	Ratio of sites
Early Neolithic	30	14
Middle Neolithic	86	39
Late Neolithic	52	24
Neolithic	11	5
Undated	12	5
Multiperiodic	24	11
Other periods	4	2
Total	219	100

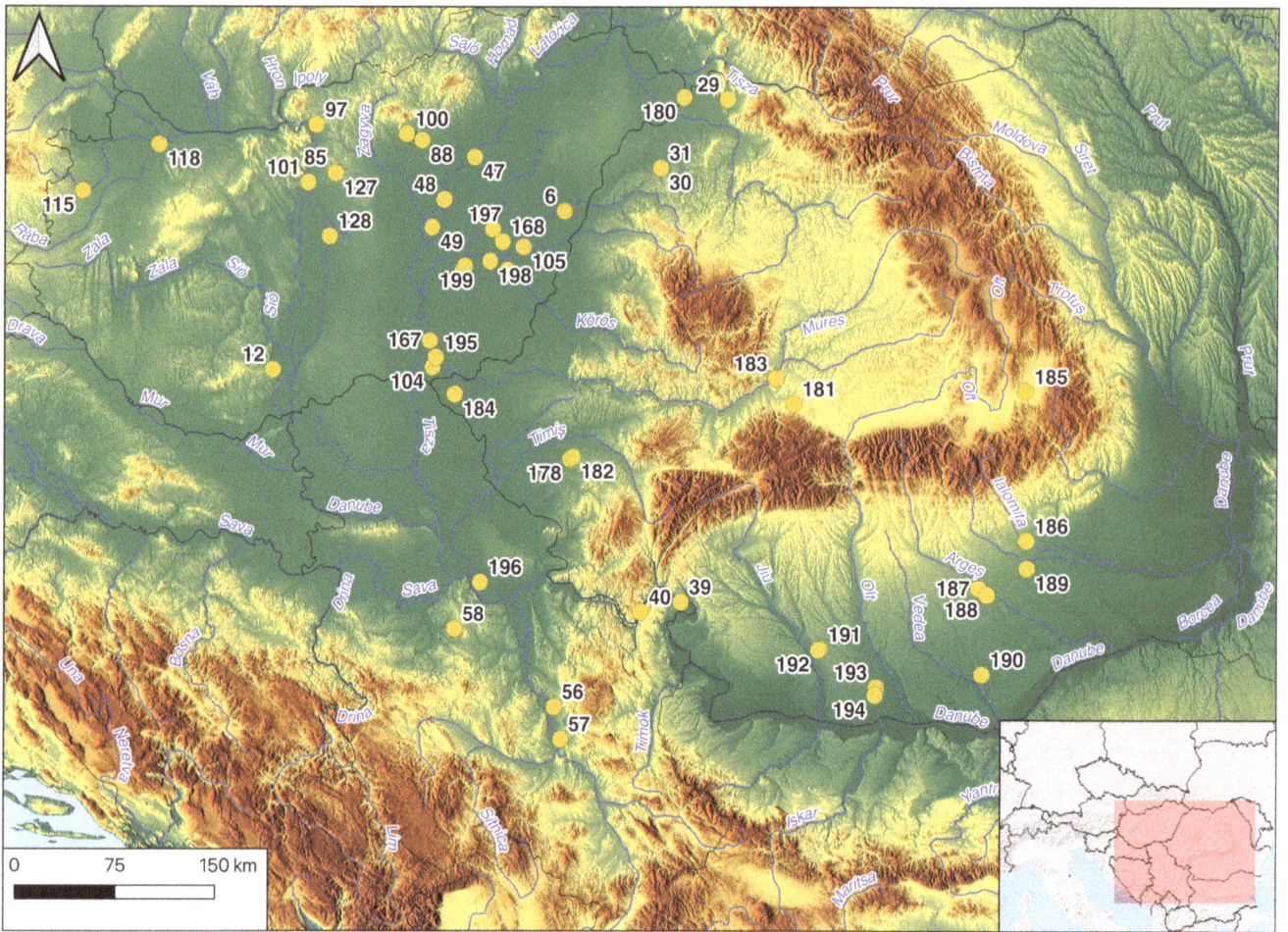

Figure 10. Obsidian distribution in the Early Neolithic (6000–5500/5400 BC)/in the first half of the 7th Millennium BC.
(Legend of the marked sites on the map: 6: Váncsod, 12: Alsónyék, 29: Călineşti-Oaş/Kányaháza, 30: Tăşnad/Tasnád, 31: Tăşnad/Tasnád, 39: Schela Cladovei, 40: Cuina Turcului, 45: Szarvas, 47: Tiszaszőlős, 48: Kőtelek, 49: Szolnok, 56: Supska, 57: Drenovac, 58: Slatina, 85: Pécel, 88: Nagyút, 90: Mezőberény, 97: Vác, 100: Visonta, 101: Budapest, 104: Deszk, 105: Szeghalom, 115: Gór, 118: Ikrény, 127: Pécel, 128: Kunpeszér, 167: Hódmezővásárhely, 168: Dévaványa, 178: Ibrány, 180: Méhtelek, 181: Miercurea Sibiului/Szerdhely, 182: Silagiu/Nagyszilas, 183: Şeuşa, 184: Dudeştii Vechi, 185: Leţ/Lécfalva, 186: Târgşoru Vechi, 187: Uliești, 188: Corbii Mari, 189: Butimanu, 190: Măgura, 191: Cârcea, 192: Cârcea, 193: Vlădila, 194: Grădinile, 195: Maroslele, 196: Starčevo, 197: Ecsegfalva, 198: Endrőd, 199: Szarvas)

In the first half of the 6th millennium BC, the appearance of Carpathian obsidian looks entirely like a mosaic. It is found along the main rivers, the Tisza and the Danube, and their side rivers, thus indicating that it was along these waterways that the material was transported (Figure 10). The Upper and Middle Tisza Region, with the Körös river system, display some minor accumulation, and the assemblages here also have a higher percentage of obsidian. Tiszaszőlős-Domaháza puszta (62%) (Domboróczki 2010: 157; Kaczanowska and Kozłowski 2012), Ibrány-Nagyerdő (87%) (Kaczanowska and Kozłowski 2010: 255), and Ecsegfalva (60%) (Kaczanowska and Kozłowski 2012: 163) have higher amounts of obsidian. We have to take into account that these sites were part of several research projects; thus, a more precise analysis (soil flotation) was applied, and this modified the rates of obsidian recognition. Still, obsidian represents less than 10% of lithic assemblages in Vojvodina and Transylvania.

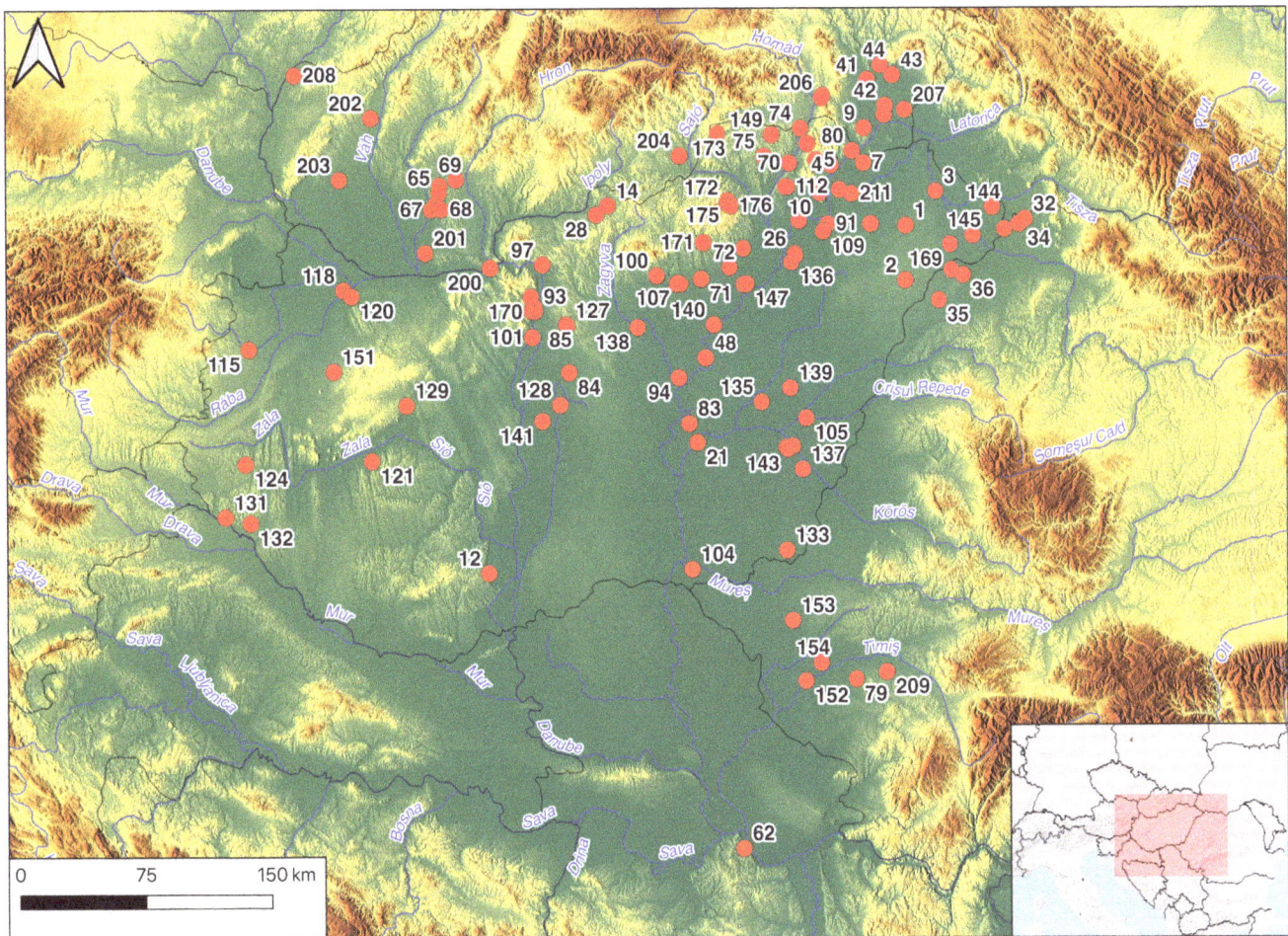

Figure 11. Obsidian distribution in the Middle Neolithic (5500/5400–5000 BC) in the second half of the 6th Millennium BC. (Legend of the marked sites on the map: 1: Besenyőd, 2: Nyírlugos, 3: Kisvarsány, 4: Erdőhorváti, 5: Baskó, 7: Bodroghalom, 9: Kašov/Kásó, 10: Bodrogkeresztúr, 12: Alsónyék, 14: Karancsság, 21: Öcsöd, 26: Polgár, 28: Szécsény, 32: Halmeu/Halmi, 34: Halmeu/Halmi, 35: Pișcolt/ Piskolt, 36: Căpleni/Kaplony, 37: Urziceni/Csanálos, 41: Moravany, 42: Slavkovce/Szalók, 43: Zalužice/Zalacska, 44: Zbudza/Izbugya, 46: Füzesabony, 48: Kőtelek, 62: Vinča-Belo Brdo, 63: Vráble/Verebély, 64: Vráble/Verebély, 65: Zitavce/Zsitvagyarmat, 66: Vlkas/Valkház, 67: Ulany/Úľany nad Žitavou/Zsitvafödémes, 68: Podhájska/Bellegszencse, 69: Mochovce, 70: Encs, 71: Füzesabony, 72: Mezőköved, 74: Hidasnémeti, 75: Felsővadász, 76: Kompolt, 79: Sátoraljaújhely, 80: Sátoraljaújhely, 83: Tiszaföldvár, 84: Inárcs, 85: Pécel, 88: Nagyút, 90: Mezőberény, 91: Tiszalök, 92: Budapest, 93: Budapest, 94: Szolnok, 95: Balsa, 97: Vác, 100: Visonta, 101: Budapest, 104: Deszk, 105: Szeghalom, 107: Kompolt, 109: Tiszavasvári, 110: Tiszavalk, 112: Megyaszó, 113: Megyaszó, 115: Gór, 118: Ikrény, 120: Ménfőcsanak, 121: Balatonszemes, 124: Gellénháza, 127: Pécel, 128: Kunpeszér, 129: Litér, 131: Muraszemenye, 132: Petrivente, 133: Battonya, 135: Dévaványa, 136: Folyás [Polgár], 137: Gerla, 138: Jászberény, 139: Kertészsziget, 140: Kisköre, 141: Kunszentmiklós [Tass], 142: Mezőberény, 143: Mezőberény, 144: Sonkád, 145: Szamossályi, 146: Tiszadob, 147: Tiszavalk, 148: Hejce, 149: Krasznokvajda, 150: Békásmegyer, 151: Kup, 152: Parța, 153: Satchinez, 154: Fratelia, 169: Nagyecsed, 170: Pomáz, 171: Eger, 172: Parasznya, 173: Aggtelek, 174: Miskolc, 175: Parasznya, 176: Miskolc, 177: Bükkárbány, 179: Polgár, 180: Méhtelek, 200: Štúrovo/Párkány, 201: Hurbanovo, 202: Borovce, 203: Veľký Grob, 204: Barca, 205: Blažice, 206: Bohdanovce, 207: Čierne Pole, 208: Kopčany, 209: Michalovce, 210: Nyírpazony, 211: Tiszabercel, 212: Veľké Raškovce)

In the second half of the 6th millennium BC, obsidian distribution was entirely different from the previous period; there was a massive accumulation in the Upper Tisza Region and the Carpathian Foothills (Figure 11). In the southern part of the Slovakian upland, most of the settlement assemblages contain more than 50% of obsidian, except Štúrovo/Párkány (41%) (Kaczanowska 1985: 26) and Bohovce (15%) (Kaczanowska 1985: 26). With the exception of the Partium, Maramures Region, the same data quality could not be collected from this region in Romania, so it is impossible to say if, or to what extent, the obsidian was transported to the east.

More than half of the entire collected Neolithic sites (n = 97) date to this period and are associated to the LBK. At this time, not just the number of sites, but also the density of obsidian in the assemblages reached a peak. Besides the domestic context, all the obsidian deposits as discussed above—with three exceptions—represent this period. Those deposits where the original archaeological context is known (e.g., Kisvarsány and Bodroghalom) were deposited in mundane rubbish pits in the settlement (Korek 1983). The quantity of obsidian decreases from the Middle Tisza Region toward the south.

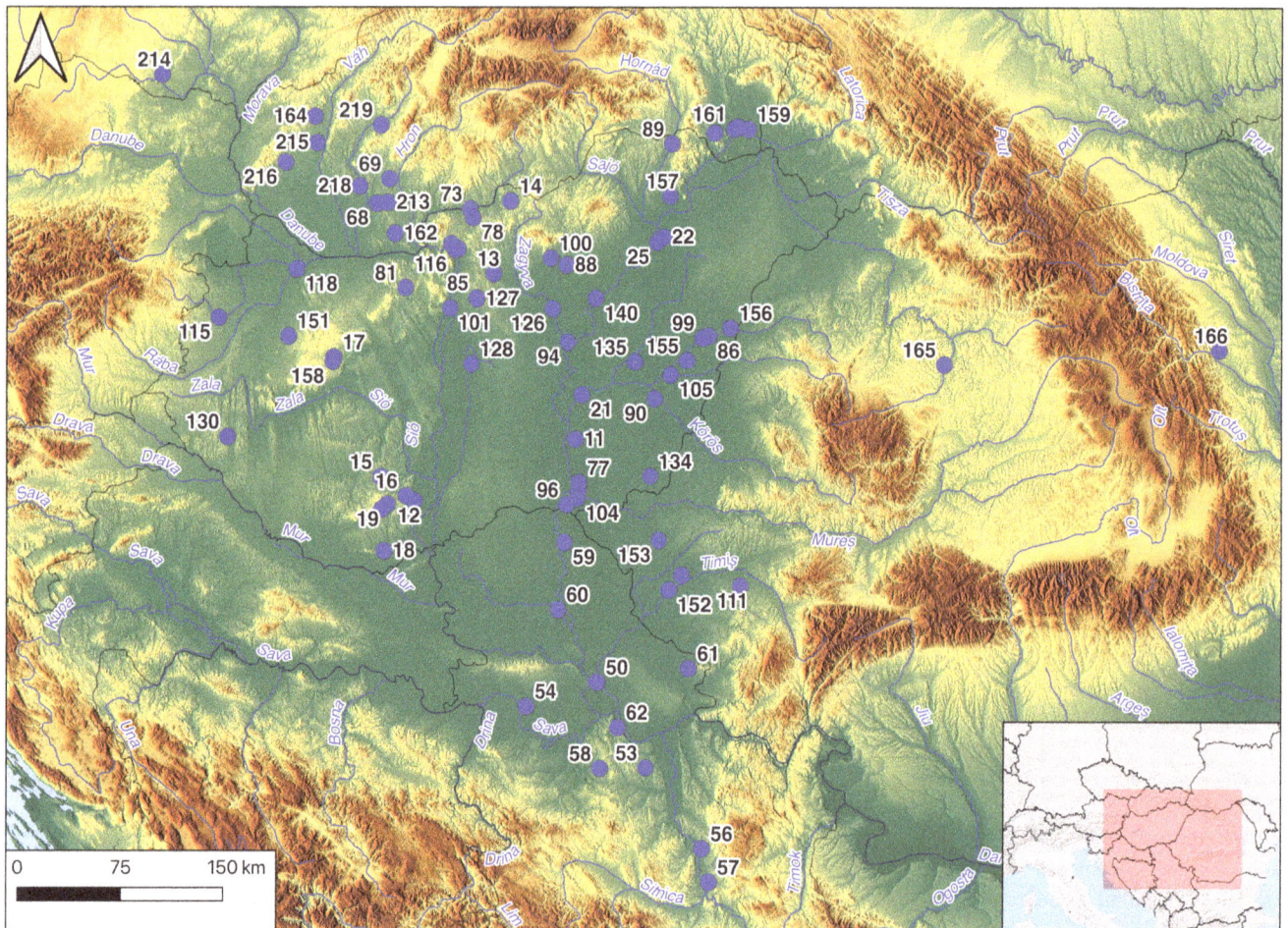

Figure 12. Obsidian distribution in the Late Neolithic (5000–4500/4400 BC) in the first half of the 5th Millennium BC. (Legend of the marked sites on the map: 11: Szegvár, 12: Alsónyék, 13: Aszód, 14: Karancsság, 15: Lengyel, 16: Mórágy, 17: Veszprém, 18: Villánykövesd, 19: Zengővárkony, 20: Pécsvárad, 21: Öcsöd, 22: Polgár, 23: Polgár, 24: Tápé, 25: Polgár, 27: Polgár, 50: Opovo, 51: Potporanj, 53: Selevac, 54: Gomolava, 55: Banjica, 56: Supska, 57: Drenovac, 58: Slatina, 59: Čoka/Csóka, 60: Borjas, 61: Vrsac, 62: Vinča-Belo Brdo, 68: Podhájska/Bellegszencse, 69: Mochovce, 73: Kolary/Koláre, 77: Hódmezővásárhely, 78: Csesztve, 81: Csabdi, 85: Pécel, 86: Berettyóújfalu, 88: Nagyút, 89: Gönc, 90: Mezőberény, 94: Szolnok, 96: Tiszasziget, 97: Vác, 99: Berettyóújfalu, 100: Visonta, 101: Budapest, 104: Deszk, 105: Szeghalom, 111: Ižkovce, 115: Gór, 116: Vác, 118: Ikrény, 123: Verőcemaros, 126: Alattyán, 127: Pécel, 128: Kunpeszér, 130: Zalaszentbalázs, 134: Battonya, 135: Dévaványa, 140: Kisköre, 151: Kup, 152: Parţa, 153: Satchinez, 154: Fratelia, 155: Darvas, 156: Esztár, 157: Szerencs, 158: Veszprém, 159: Čičarovce, 160: Hrčeľ, 161: Oborín, 162: Svodín/Szőgyén, 163: Svodín/Szőgyén, 164: Žlkovce, 165: Iclod, 166: Tîrpeşti, 213: Bardoňovo, 214: Těšetice-Kyjovice, 215: Pečeňady, 216: Budmerice, 217: Veľké Raškovce/Nagyráska, 218: Branč, 219: Brodzany)

In the first half of the 5th millennium BC, the spatial and quantitative distribution of obsidian changed. The Upper Tisza Region concentration disappeared, while the eastern part of the Slovakian lowlands in the river valleys displays more prominent roles in the obsidian transfer (Tóth et al. 2011) (Figure 12). Kolary/Koláre (80%) (Kaczanowska 1985: 82; Biró 2014: 55, Table 1a) and Veľké Raškovce/Nagyráska (62%) (Kaczanowska 1985: 126); otherwise, the ratio is less than 15% in other cases. In the Hungarian territory, only the sites of Mezőberény-Bódishát (36%) (Biró 1998, 2014: 55, Table 1a) and Csesztve-Stalák (31%) (Biró 2014: 55, Table 1a) had more than 30% obsidian in the lithic assemblages. Along the Tisza River, the southern part of the Carpathian Basin became more important concerning obsidian distribution. The Tisza and Danube rivers probably served as natural corridors in the Late Neolithic communication and exchange system. Besides, the Danube in the Vršac region shows some kind of accumulation in obsidian exchange, based on the Vršac-At, Opovo, Perlez, Potporanj-Kremenjak, and Potporanjska granica sites (Milić 2016: 320; Marić 2015: 44–45). There is no evidence of obsidian in the territory between the Mur, Drava, and the Sava; the Late Neolithic grave from Gomolava is one exception (Biró 1998; Kaczanowska and Kozłowski 1986). The Adriatic region's entire Neolithic does not show any connection with the circulation of Carpathian obsidian.

Discussion

To study the economic value of obsidian and its role in the Neolithic value system, I would like to highlight the community level, that is, the social context where the concept of value was formed and linked with social and ritual content (Peterson et al. 1997). Value is a subjective concept determined by social interaction in real-life contexts and thus variable and community specific. Nevertheless, it is crucial to have an idea of how, in what ways, and which values and value systems governed prehistoric societies. There is a large amount of philosophical, anthropological, and economic literature on value (Sahlins 1972; Wittgenstein 1980; Kopytoff 1986; Munn 1992; Graeber 2001; Appadurai 2006; Graeber 2014; Gregory 2015). For prehistoric archaeology, however, a practical approach needs to be built upon archaeological objects and their contexts. Thus, while it is a complex and overarching research topic, I will focus on the practices involving humans and obsidian items on a community level.

The different LBK and Lengyel communities did not have the same connections to obsidian, depending on how far they lived from the source. We have seen examples from Szécsény, Karancsság, and other sites also in the Northern Hungarian Mid-Mountains where the local communities had a high interest in obsidian insofar as it dominated their lithic raw material use, besides the local (limnic silicite) material. Neolithic communities in the South-Transdanubian region (e.g., Alsónyék, Zengővárkony, Lengyel, Villánykövesd) were located far away from the obsidian source, and people seem to have had less interest. While they could have obtained obsidian via their connections with other LBK or Lengyel communities who had direct or indirect access, they did not pursue these possibilities to any larger extent. The non-obsidian, local lithics clearly dominated, and obsidian was rare in the domestic and burial context in South-Transdanubian assemblages. All in all, southern Hungarian sites show that obsidian did not stand out in all contexts in which it appeared; when it came to practical use and depositional practices, it was replaceable by other raw materials—a fact that should lead us to de-emphasize the economic importance of the obsidian.

The practicalities of different raw materials, which are likely connected to their economic value, can be characterized with respect to their scarcity, the investment needed for procurement, and the technologies required to produce an item. Beyond the economic value, the recognition of a lithic's social value is more complex, and almost every region and community with a diverse landscape could have different ways of treating and valuing lithics. I define social value as how things are integrated into social relations and their role in interactions between and within communities. Social value is determined by the social context in which it is used, and it can be described along two main axes, the vertical of differential social power within a community (social inequality) and the horizontal social differences within and between communities.

This research has focused on the archaeological context of obsidian and compared it with other lithics. The grave goods give us an extraordinary opportunity to see patterns of selection strategy, habits of positioning items in a grave, and the selection of tool types used by different communities to express their social status, or at least their social role in the burial practice. Obsidian in burials did not differ in their positioning, type, and quantity from other lithic raw materials (Figure 5). Obsidian was deposited as a blade and core beside the deceased in the second half of the 6th millennium and the first half of the 5th millennium BC. There are just a few exceptions, such as the late LBK burial from Polgár-Ferenci hát (Grave 867/1230, and 486/687; the burial activity is dated between 5300 and 5070 cal BC) with an extra-large blade core (Whittle et al. 2013: 73–75; Lipson et al. 2017). In the first half of the 5th millennium BC at Alsónyék-Bátaszék, 6 obsidian cores, 5 flakes, and 25 blades were discovered as grave goods in 32 Lengyel burials, which is an exceptional case (Szilágyi 2019a: 293–294).

In summary, obsidian did not play a significant role in burial ritual, and there is nothing to indicate a connection with an elevated status of the deceased or the buried community. Instead, we could focus on the visual qualities of the obsidian and its esthetic beauty, which may have been important at that time. On the one hand, it was an easily recognizable material, and its high quality made it crucial for toolmaking, which probably created more intensive demand for it. On the other hand, the knappability and interest in obsidian made it an easily exchangeable material, which created connectivity between communities. This correlates with the archaeological evidence that some regions (e.g., the Vršac area), although far from the geologic source, have shown higher interest in obsidian than others in their surroundings.

Conclusion

The domestic, burial, and ritual contexts of obsidian deposition in the Carpathian Basin show us that there were probably not any universal set of rules that directly related to this specific raw material. Indeed, obsidian was widely distributed across space, but more or less related to the Tisza River and its tributaries. From a temporal perspective, the peak of obsidian's circulation is restricted to the second half of the 6th millennium and first half of the 5th millennium BC. The burial and ritual depositions of obsidian were homogeneous, expressed in a precise position inside the grave in core or blade form. This suggests the possibility of a specific intention behind this visual appearance and action during burial. All three depositional contexts seem to show the varying behavior of different communities, indicating that the obsidian could have had different values.

The Neolithic value systems (economic, social, and ritual) were diverse and multidimensional: one item or raw material could have simultaneously had different kinds of value. In identifying what types of value played an essential role in the last usage or the final phase of the object biography, we have to pay attention to the exact depositional (archaeological) context. The diversity of spatial, temporal, and social contexts that influence an object's value was probably highly localized, as we can see with the current example. Obsidian was a widely distributed material—with a long research history—but the peak of its distribution was restricted to some regions (Upper-Middle Tisza Region and the Slovakian uplands) and some centuries (5500/5400–4500 BC). The specific depositional pattern (nodules, precores, and blade cores) is located around the Tokaj-Eperjes Mountain area.

References

Allard, Pierre, Laurent Klaric, and Bibiana Hromadová
  2017 Obsidian Blade Debitage at Kašov-Čepegov I (Bükk Culture), Slovakia. *Bulgarian E-Journal of Archaeology* 7: 17–35.

Appadurai, Arjun
  2006 Introduction: Commodities and the Politics of Value. In *The Social Life of Things. Commodities in Cultural Perspective*, edited by Arjun Appadurai, pp. 3–63. Cambridge University Press, Cambridge.

Bačo, Pavel, Ľubomíra Kaminská, Jaroslav Lexa, Zoltán Pécskay, Zuzana Bačová, and Vlastimil Konečny
  2017 Occurrences of Neogene Volcanic Glass in the Eastern Slovakia – Raw Material Source for the Stone Industry. *Anthropologie* LV(1–2): 207–230.

Bačo, Pavel, Jaroslav Lexa, Zuzana Bačova, Patrik Konečný, and Zoltán Pécskay
  2018 Geological Background of the Occurrences of Carpathian Volcanic Glass, Mainly Obsidian, in Eastern Slovakia. *Archeometriai Műhely* XV(3): 157–166.

Bácsmegi, Gábor
  2003 A Lengyeli Kultúra Temetkezései Karancsságon. *Móra Ferenc Múzeum Évkönyvei, Studia Archaeologica* IX: 81–86.
  2014a Geoarceológiai és környezettörténeti kutatások Karancsság-Alsó-Rétek lelőhelyen. PhD dissertation, Szeged, Szegedi Tudományegyetem.
  2014b Geoarcheological and Environmental Historical Studies on the Karancsság-Alsó-Rétek site. PhD dissertation (short English summary), Szeged, Szegedi Tudományegyetem.

Bajnóczi, Bernadett, Gabriella Schöll-Barna, Nándor Kalicz, Zsuzsanna Siklósi, George H. Hourmouziadis, Fotis Ifantidis, Aikaterini Kyparissi-Apostolika, Maria Pappa, Rena Veropoulidou, and Christina Ziota
  2013 Tracing the Source of Late Neolithic Spondylus Shell Ornaments by Stable Isotope Geochemistry and Cathodoluminescence Microscopy. *Journal of Archaeological Science* 40(2): 874–882.

Bánesz, Ladislav
  1974 Hromadný Nález Obsidiánovej Suroviny Na Gravettskom Sídlisku v Cejkove, Okr. Trebišov. *Archeologické Rozhledy* 26: 51–54.
  1991 Neolitická Dielňa Na Výrobu Obsidiánovej Industrie v Kašove. *Východoslovenský Pravek* 3: 39–68.
  1993 Neolitische Werkstatt zur Herstellung von Obsidianindustrie in Kašov. In *Actes du XIIe Congrès international des sciences préhistoriques et protohistoriques: Bratislava, 1–7 septembre 1991*, edited by Juraj Pavúk, Marián Fabis, Ivan Kuzma,

Klára Marková, Ladislav Bánesz, Peter Bednár, Darina Bialeková, Gabriel
Fusek, and Danica Stassíková-Stukovská, pp. 432–437. Institut d'archéologie de
l'Académie slovaque des sciences, Union internationale des sciences préhistoriques
et protohistoriques, Bratislava, Nitra, Slovakia.

Biagi, Paolo, and Elisabetta Starnini
2013   Pre-Balkan Platform Flint in the Early Neolithic Sites of the Carpathian
Basin: Its Occurrence and Significance. In *Moments in Time. Papers Presented
to Pál Raczky on His 60th Birthday*, edited by Alexandra Anders, Gábor Kalla,
Viktória Kiss, Gabriella Kulcsár, and Gábor V. Szabó, pp. 35–46. Ősrégészeti
Tanulmányok, I. L'Harmattan, Budapest.

Biró, Katalin T.
1981   A Kárpát-medencei obszidiánok vizsgálata. *Archaeologiai* Értesítő
108: 196–205.
1984   Őskőkori és kőkori pattintott kőeszközeink nyersanyagának forrásai.
Archaeologiai Értesítő 111: 42–52.
1987   Chipped Stone Industry of the Linearband Pottery Culture in Hungary. In
Chipped Stone Industries of the Early Farming Cultures in Europe, edited by Janusz K.
Kozłowski and Stefan K. Kozłowski, pp. 131–167. Archaeologia Interregionalis.
Warsaw University Press, Warsaw.
1989   A lengyeli kultúra dél-dunántúli kőeszköz-leletanyagainak nyersanyagáról
I. *Communicationes Archaeologicae Hungariae* 41: 22–31.
1990   A lengyeli kultúra dél-dunántúli kőeszköz-leletanyagainak nyersanyagáról
II. *Communicationes Archaeologicae Hungariae* 42: 66–76.
1998   *Lithic Implements and the Circulation of the Raw Materials in the
Great Hungarian Plain during the Late Neolithic Period*. Magyar Nemzeti
Múzeum, Budapest.
2004   A kárpáti obszidiánok: legenda és valóság. *Archeometriai Műhely* 1(1): 3–8.
2009   Egy sváb menyecske hozománya. Gondolatok a szegvár-tűzkövesi
kőeszköz raktárlelet kapcsán. In *Medinától Etéig. Régészeti tanulmányok Csalog József
születésének 100. évfordulójára*, edited by Bende Lívia and Lőrinczy Gábor, pp.
103–115. Móra Ferenc Múzeum, Szentes, Hungary.
2014   Carpathian Obsidians: State of Art. In *Lithic Raw Material Exploitation and
Circulation in Prehistory: A Comparative Perspective in Diverse Palaeoenvironments*,
edited by Masayoshi Yamada and Akira Ono, pp. 47–68. Etudes et Recherches
Archéologiques de l'Université de Liège 18. Université de Liège, Service de
préhistoire & Centre de recherches archéologiques, Liège.

Biró, Katalin T., Zsolt Kasztovszky, and Andrea Mester
2021   New-Old Obsidian Nucleus Depot Find from Besenyőd, NE Hungary.
In *Beyond the Glass Mountains. Papers Presented for the 2019 International Obsidian
Conference*, May 27–29, 2019, Sárospatak, edited by Katalin T. Biró and András
Markó, pp. 95–108. Inventaria Praehistorica Hungariae, XIV. Magyar Nemzeti
Múzeum, Budapest.

Burgert, Pavel, Antonín Přichystal, Lubomír Prokeš, Jan Petřík, and Somina Hušková
 2017   The Origin and Distribution of Obsidian in Prehistoric Bohemia.
 Bulgarian E-Journal of Archaeology 7: 1–15.

Cheben, Michal, Pavla Hrselová, Maria Wunderlich, and Kata Szilágyi
 2020   Stone Tools from the LBK and Želiezovce Settlement Site of Vráble. In
 Archaeology in the Žitava Valley I. The LBK and Želiezovce Settlement Site of Vráble,
 edited by Martin Furholt, Ivan Cheben, Johannes Müller, Alena Bistáková, Maria
 Wunderlich, and Müller-Scheeßel, Nils, pp. 363–416. Sidestone Press, Leiden.

Domboróczki, László
 2010   Report on the Excavation at Tiszaszőlős-Domaháza-Puszta and a New
 Model for the Spread of the Körös Culture. In *Neolithization of the Carpathian
 Basin: Northernmost Distribution of the Starčevo/Körös Culture; Papers Presented on the
 Symposium Organized by the EU Project FEPRE (The Formation of Europe: Prehistoric
 Population Dynamics and the Roots of Socio-Cultural Diversity)*, edited by Janusz
 K. Kozłowski and Pál Raczky, pp. 137–176. Polska Akademia Umiejętności,
 Kraków-Budapest.

Fábián, Szilvia
 2005   Arcos edénytöredékek a zselizi kultúra lelőhelyéről, Szécsény-Ültetésről.
 Archaeologiai Értesítő 130(1–2): 5–20.
 2010.   Siedlung Der Zseliz-Periode Der Linearbandkeramik in Szécsény. *Antaeus*
 31–32: 225–283.
 2012   Szécsény-Ültetés Újkőkori Településének Időrendje Az Import
 Kerámialeletek Tükrében. In MΩMOΣ *III. Őskoros Kutatók IV.* Összejövetelének
 Konferenciakötete, edited by Emese Gyöngyvér Nagy, pp. 191–213. Déri Múzeum,
 Debrecen, Hungary.

Fábián, Szilvia, Roderick B. Salisbury, Gábor Serlegi, Nicklas Larsson, Szilvia Guba,
and Gábor Bácsmegi.
 2016   Early Settlement and Trade in the Ipoly Region: Introducing the Ipoly-
 Szécsény Archaeological Project. *Hungarian Archaeology E-Journal* 2 (Spring): 1–10.

Faragó, Norbert, Zsolt Mester, and Attila Király
 2019   The Knapped Stone Assemblage from Boldogkőváralja in the Light of a
 New Statistical Evaluation. *Litikum* 7–8: 55.

Furholt, Martin, Jozef Bátora, Ivan Cheben, Helmut Kroll, and Peter Tóth
 2014   Vráble-Velké Lehemby: Eine Siedlungsgruppe Der Linearkeramik in Der
 Südwestslowakei. Vorbericht Über Die Untersuchungen Der Jahre 2010 Und
 2012 Und Deutungsansätze. *Slovenská Archeológia* 62(2): 227–266.

Furholt, Martin, Ivan Cheben, Johannes Müller, Alena Bistáková, Maria Wunderlich, and Müller-Scheeßel, Nils (editors)

    2020   *Archaeology in the* Žitava *Valley I. The LBK and* Želiezovce *Settlement Site of Vráble.* Vol. 9. Scales of Transformation. Sidestone Press, Leiden.

Gatsov, Ivan, and Petranka Nedelcheva

    2015   Flint Caches in the Eneolithic Settlement Pietrele-Măgura Gorgana, Romania. In *Neolithic and Copper Age between the Carpathians and the Aegean Sea: Chronologies and Technologies from the 6th to the 4th Millennium BCE: International Workshop Budapest 2012*, edited by Svend Hansen, Pál Raczky, Alexandra Anders, and Agathe Reingruber, pp. 295–301. Archäologie in Eurasien 31. Bonn: Habelt.

Graeber, David

    2001   *Toward an Anthropological Theory of Value: The False Coin of Our Own Dreams.* Palgrave Macmillan, New York.
    2014   *Debt – Updated and Expanded: The First 5,000 Years.* Melville House, Brooklyn.

Gregory, Christopher A.

    2015   *Gifts and Commodities.* Second. Hau Books, Chicago.

Hillebrand, Jenő

    1928   A Nyírlugosi Obsidiannucleus Depotleletről [On the Nyírlugos Obsidian Core Depot Find]. *Archaeologiai* Értesítő 42: 39–42.

Inizan, Marie-Louise, Michèle Reduron-Ballinger, Helene Roche, and Jacques Tixier

    1999   *Technology and Terminology of Knapped Stone.* Translated by J. Féblot-Augustins. Préhistoire de La Pierre Taillée. Cercle de Recherches et d'Etudes Prehistoriques, Nanterre, France.

Kaczanowska, Małgorzata

    1985   *Rochstoffe, Technik und Typologie der neolithischen Feuersteinindustrien im Nordtei des Flussgebietes der Mitteldonau.* Panstwowe Wydawnictwo Naukowe, Warsawa.

Kaczanowska, Małgorzata, and Janusz K. Kozłowski

    1986   Frühlengyel-Feuersteinindustrie von Svodín. In *Internationales Symposium über die Lengyel-Kultur. Nové Vozokany 5–9. November 1984*, edited by Bohuslav Chropovský, pp. 121–133. Archäologisches Institut der Slowakischen Akademie der Wissenschaften, Nitra, Vienna.
    2010   Chipped Stone Industry from Ibrány. In *Neolithization of the Carpathian Basin: Northernmost Distribution of the Starčevo/Körös Culture; Papers Presented on the Symposium Organized by the EU Project FEPRE (The Formation of Europe: Prehistoric Population Dynamics and the Roots of Socio-Cultural Diversity)*, edited by Janusz

K. Kozłowski and Pál Raczky, pp. 254–265. Polska Akademia Umiejętności, Kraków-Budapest.

2012   Körös Lithics. In *The First Neolithic Sites in Central/South-East European Transect*, edited by Alexandra Anders and Zsuzsanna Siklósi, pp. 161–170. British Archaeological Reports International Series 2334. Archaeopress, Oxford.

Kaczanowska, Małgorzata, Janusz K. Kozłowski, and Stanislav Šiška
1993   *Neolithic and Eneolithic Chipped Stone Industries from Šarišské Michaľany, Eastern Slovakia: Linear Pottery, Bükk and Baden Cultures*. Kraków: Institute of Archaeology, Jagellonian University, Kraków.

Kalicz, Nándor
2011   *Méhtelek: The First Excavated Site of the Méhtelek Group of the Early Neolithic Körös Culture in the Carpathian Basin*. BAR International Series 2321. Archaeopress, Oxford.

Kaminská, Ľubomíra.
2013   Sources of Raw Materials and Their Use in the Palaeolithic of Slovakia. In *The Lithic Raw Material Sources and Interregional Human Contacts in the Northern Carpathian Regions*, edited by Zsolt Mester, pp. 99–109. Polish Academy of Arts and Sciences Kraków, Institute of Archaeological Sciences of the Eötvös Loránd University Budapest, Kraków and Budapest.

2018   Use of Obsidian from the Paleolithic to the Bronze Age in Slovakia. *Archeometriai Műhely* XV(3): 197–212.

Kasztovszky Zsolt, and Katalin T. Bíró
2004   A kárpáti obszidiánok osztályozása promt gamma aktivizációs analízis segítségével: geológiai és régészeti mintákra vonatkozó első eredmények. *Archeometriai Műhely* 1(1): 9–15.

Kasztovszky, Zsolt, and Antonín Přichystal
2018   An Overview of the Analytical Techniques Applied to Study the Carpathian Obsidians. *Archeometriai Műhely* XV(3): 187–196.

Kohút, Milan, John A. Westgate, Nicholas J. G. Pearce, and Pavel Bačo
2021   The Carpathian Obsidians – Contribution to Their FT Dating and Provenance (Zemplín, Slovakia). *Journal of Archaeological Science: Reports* 37: 102861.

Kopytoff, Igor
1986   The Cultural Biography of Things: Commoditization as Process. In *The Social Life of Things: Commodities in Cultural Perspective*, edited by Arjun Appadurai, pp. 64–94. Cambridge University Press, Cambridge.

Korek, József
    1983   Adatok a Tiszahát Neolitikumához. *Jósa András Múzeum* Évkönyve
    1975–1977 (20): 8–60.

Kovács, Katalin
    2013   Late Neolithic Exchange Networks in the Carpathian Basin. In *Moments
    in Time. Papers Presented to Pál Raczky on His 60th Birthday*, edited by Alexandra
    Anders, Gábor Kalla, Viktória Kiss, Gabriella Kulcsár, and Gábor V. Szabó, pp.
    385–400. Ősrégészeti Tanulmányok, I. L'Harmattan, Budapest.

Lexa, Jaroslav, Ioan Seghedi, Karoly Németh, Alexandru Szakács, Vlastimil Konečny,
Zoltan Pécskay, Alexandrina Fülöp, and Marinel Kovacs
    2010   Neogene-Quaternary Volcanic Forms in the Carpathian-Pannonian
    Region: A Review. *Open Geosciences* 2(3): 207–270.

Lipson, Mark, Anna Szécsényi-Nagy, Swapan Mallick, Annamária Pósa, Balázs Stég-
már, Victoria Keerl, Nadin Rohland, et al.
    2017   Parallel Palaeogenomic Transects Reveal Complex Genetic History of
    Early European Farmers. *Nature* 551(7680): 368–372.

Marić, Miroslav
    2015   Modelling Obsidian Trade Routes during Late Neolithic in the South-East
    Banat Region of Vršac Using GIS. *Starinar* 65: 37–52.

Mester, Zsolt, and Jacques Tixier
    2013   Pot à Lames: The Neolithic Blade Depot from Boldogkőváralja
    (Northeast Hungary). In *Moments in Time. Papers Presented to Pál Raczky on His
    60th Birthday*, edited by Alexandra Anders, Gábor Kalla, Viktória Kiss, Gabriella
    Kulcsár, and Gábor V. Szabó, pp. 173–186. Ősrégészeti Tanulmányok, I.
    L'Harmattan, Budapest.

Milić, Marina
    2016   Obsidian Exchange and Societies in the Balkans and the Aegean
    from the Late 7th to 5th Millennia BC. PhD dissertation, University College
    London, London.

Müller-Scheeßel, Nils, Ivan Cheben, Dragana Filipović, Zuzana Hukeľová, and
Martin Furholt
    2016   The LBK Site of Vráble in Southwestern Slovakia: Selected Results of
    the Excavation Season 2016. In *Bericht der Römisch-Germanischen Kommission*, pp.
    81–128. Deutsches Archäologisches Institut. Gebr Mann Verlag, Berlin.

Müller-Scheeßel, Nils, and Zuzana Hukeľová
    2020   The Burials and Human Remains from the LBK and Želiezovce Settlement

Site of Vráble. In *Archaeology in the Žitava Valley I. The LBK and Želiezovce Settlement Site of Vráble*, edited by Martin Furholt, Ivan Cheben, Johannes Müller, Alena Bistáková, Maria Wunderlich, and Nils Müller-Scheeßel, pp. 159–236. Scales of Transformation 9. Sidestone Press, Leiden.

Munn, Nancy D.
1992   *The Fame of Gawa: A Symbolic Study of Value Transformation in a (Papua New Guinea) Massim Society.* Duke University Press, Durham, North Carolina.

Nandris, John
1975   A Re-Consideration of the South-East European Sources of Archaeological Obsidian. *Univ. of London Bull. of the Inst. of Archaeology* 12: 71–94.

Odell, Georg H.
2006   *Lithic Analysis.* University of Tulsa Press, Oklahoma.

Patay Róbert, and Marton Tibor
2018   Középső neolitikus kőeszköz raktárlelet Kosdról. Presented at the IX. Kőkor Kerekasztal Konferencia, December 7, Szeged.

Pécskay, Zoltán, and Ferenc Molnár
2002   Relationship between Volcanism and Hydrothermal Activity in the Tokaj Mountains, Northeast Hungary, Based on K-AR Ages. *Geologica Carpathica* 53(5): 303–314.

Pécskay, Zoltán, Jaroslav Lexa, Alexandru Szakács, Ioan Seghedi, Kadosa Balogh, Vlastimil Konečný, Tibor Zelenka, et al.
2006   Geochronology of Neogene Magmatism in the Carpathian Arc and Intra-Carpathian Area. *Geologica Carpathica* 57(6): 511–530.

Pécskay, Zoltán, Kadosa Balogh, Vilma Székyné Fux, and Pál Gyarmati
1987   A Tokaji-Hegység Miocén Vulkánosságának K/Ar Geokronológiája (K/Ar Geochronology of the Miocene Volcanism in the Tokaj Mts. *Földtani Közlöny/Bulletin of the Hungarian Geological Society* 117: 237–253.

Peterson, Jane, Douglas R. Mitchell, and M. Steven Shackley
1997   The Social and Economic Contexts of Lithic Procurement: Obsidian from Classic-Period Hohokam Sites. *American Antiquity* 62(2): 231–259.

Přichystal, Antonín, and Petr Škrdla
2014   Kde Ležel Hlavní Zdroj Obsidiánu v Pravěku Střední Evropy? *Slovenská Archeológia* LXII(2): 215–226.

Priskin, Anna, Vajk Szeverényi, and Balázs Wieszner

2019   Obsidian Exchange in Early Neolithic Eastern Hungary. In *IOC 2019*, edited by András Markó, Kata Szilágyi, and Katalin T. Biró, p. 53. Hungarian National Museum, Sárospatak.

Rácz, Béla

2008   Pattintott kőeszköz-nyersanyagok felhasználásának előzetes eredményei a paleolitikumban a mai Kárpátalja területén. *Archeometriai Műhely* 5(2): 47–54.

2009   Kárpátalja paleolit nyersanyag-felhasználási régióinak elsődleges nyersanyagai. In *MΩMΩΣ VI – Őskoros kutatók VI.* Összejövetele. *Nyersanyagok és kereskedelem*, edited by Ilon Gábor, pp. 321–326. Kulturális Örökségvédelmi Szakszolgálat, Vas megyei Múzeumok Igazgatósága, Kőszeg.

2012   Kárpátaljai obszidiánok: szakirodalmi adatok és terepi tapasztalatok. In *Környezet-Ember-Kultúra. A természettudományok* és *a régészet párbeszéde*, edited by Kreiter Attila, Pető Ákos, and Tugya Beáta, pp. 353–362. Magyar Nemzeti Múzeum Örökségvédelmi Központ, Budapest.

2013   Main Raw Materials of the Palaeolithic in Transcarpathian Ukraine: Geological and Petrographical Overview. In *The Lithic Raw Material and Interregional Human Contacts in the Northern Carpathian Regions*, edited by Zsolt Mester, pp. 131–147. Polish Academy of Arts and Sciences, Institute of Archaeological Sciences of the Eötvös Loránd University, Kraków and Budapest.

2018   The Carpathian 3 Obsidian. *Archeometriai Műhely* XV(3): 181–186.

Rhyzov, Sergey

2018   Archaeological and Geological Studies of Obsidians in Ukrainian Transcarpathia. *Archeometriai Műhely* XV(3): 225–230.

Sahlins, Marshall

1972   *Stone Age Economics.* Aldine de Gruyter, New York.

Scharl, Silviane, Solène Denis, Ingrid Koch, Daniel Schyle, Birgit Gehlen, Jean-Philippe Collin, Pierre Allard, Vincent Delvigne, and Marjorie de Grooth

2021   Studying Neolithic Lithics – From a Cross-Border Dialogue to a Common Language. *Journal of Neolithic Archaeology* 23: 115–128.

Siklósi, Zsuzsanna

2013   *Traces of Social Inequality during the Late Neolithic in the Eastern Carpathian Basin.* Dissertationes Pannonicae, Ser. IV. Vol. 3. Eötvös Loránd University, Institute of Archaeological Sciences, L'Harmattan, Budapest.

Siklósi, Zsuzsanna, and Piroska Csengeri

2011   Reconsideration of Spondylus Usage in the Middle and Late Neolithic of the Carpathian Basin. In *Spondylus in Prehistory. New Data and Approaches. Contribution to the Archaeology of Shell Technologies*, edited by Fotis Ifantidis and

Marianna Nikolaidou, pp. 47–62. British Archaeological Reports International Series 2216. Archaeopress, Oxford.

Siklósi, Zsuzsanna, Zsuzsanna M. Virág, Viktória Mozgai, and Bernadett Bajnóczi
2017   The Spread of the Products and Technology of Metallurgy in the Carpathian Basin between 5000 and 3000 BC – Current Questions. *Dissertationes Archeologicae* 3(5): 67–82.

Soós, Virág
1982   Előzetes Jelentés a Szécsény-Ültetési Zselizi Telep Feltárásától. *A Nógrád Megyei Múzeumok* Évkönyve VIII: 7–36.

Szepesi, János, Réka Lukács, Katalin T. Biró, András Markó, Zoltán Pécskay, and Szabolcs Harangi
2009   Karancsság – Alsó rétek lelőhely középső és késő neolit pattintott kőeszközei. BA szakdolgozat, Szeged: Szegedi Tudományegyetem.
2018   Geology of Tokaj Mountains Obsidians. *Archeometriai Műhely* XV(3): 167–180.

Szilágyi, Kata
2019a   A késő neolitikus lengyeli kultúra délkelet-dunántúli csoportjának pattintott kőeszközkészítő tevékenysége. PhD dissertation, Eötvös Loránd Tudományegyetem, Budapest. http://hdl.handle.net/10831/46276.
2019b   The Chipped Stone Tools Production Activity of the Late Neolithic Lengyel Culture's South-Eastern Transdanubian Group. *Archeometriai Műhely* XVI(2): 85–98.

Tóth, Peter, Peter Demján, and Kristina Griačová
2011   Adaptation of Settlement Strategies to Environmental Conditions in Southern Slovakia in the Neolithic and Eneolithic. *Documenta Praehistorica* 38: 307–321.

Tripković, Boban, and Marina Milić
2008   The Origin and Exchange of Obsidian from Vinča-Belo-Brdo. *Starinar* LVIII: 71–76.

Whittle, Alasdair, Alexandra Anders, Alexander R. Bentley, Penny Bickle, Lucy Cramp, László Domboróczki, Linda Fibiger, et al.
2013   3 Hungary. In *The First Farmers of Central Europe. Diversity in LBK Lifeways*, edited by Penny Bickle and Alasdair Whittle, pp. 49–100. Oxbow Books, Oxford, Oakville.

Williams, Olwen, and John Nandris

    1977   The Hungarian and Slovak Sources of Archaeological Obsidian: An Interim Report on Further Fieldwork, with a Note on Tektites. *Journal of Archaeological Science* 4(3): 207–219.

Williams Thorpe, Olwen, S. E. Warren, and J. G. Nandris

    1984   The Distribution and Provenance of Archaeological Obsidian in Central and Eastern Europe. *Journal of Archaeological Science* 11(3): 183–212.

Wittgenstein, Ludwig

    1980   *Culture and Value.* Edited by Georg Henrik von Wright and Heikki Nymann. Translated by Peter Winch. Basil Blackwell, Oxford.

Zalai-Gaál, István

    1986   Sozialarchäologische Forschungsmöglichkeiten aufgrund spätneolithischer Gräbergruppen in Südwestlichen Ungarn. *A Béri Balogh Ádám Múzeum* Évkönyve 14: 139–154.

    2004   Der spätneolithische geschliffene Steingerätbestand in Südtransdanubien. II. Katalog. *Wosinszky Mór Múzeum* Évkönyve 26: 83–135.

    2005   Der spätneolithische geschliffene Steingerätbestand in Südtransdanubien III. Abbildungen. *Wosinszky Mór Múzeum* Évkönyve 27: 159–204.

    2010   *Die soziale Differenzierung im Spätneolithikum Südtransdanubiens. Die Funde und Befunde aus den alten Ausgrabungen.* Varia Archaeoligica Hungarica 24. Archaeolingua, Budapest.

Zelenka, Tibor, Pál Gyarmati, and János Kiss

    2012   Paleovolcanic Reconstruction in the Tokaj Mountains. *Central European Geology* 55(1): 49–83.

CHAPTER 3

Obsidian Artifacts from Tell Hódmezővásárhely-Gorzsa (SE Hungary): Results of a Provenance Study Using pXRF

CLIVE BONSALL, ELISABETTA STARNINI, BARBARA VOYTEK, AND FERENC HORVÁTH

Clive Bonsall University of Edinburgh, UK (c.bonsall@ed.ac.uk)

Elisabetta Starnini University of Pisa, Italy (elisabetta.starnini@unipi.it)

Barbara Voytek University of California, Berkeley, USA (bvoytek@berkeley.edu)

Ferenc Horváth Móra Ferenc Museum, Szeged, Hungary (f_horvath@mfm.u-szeged.hu)

Abstract

In total, 175 obsidian artifacts from Late Neolithic (Tisza culture) contexts at the tell site of Gorzsa in southeast Hungary were analyzed using a portable XRF device and the results were compared with the corresponding measurements made on geological samples from known European obsidian sources. The data support the conclusion that most of the obsidian used at Gorzsa originated in the Carpathian 1 (C1 – Cejkov-Viničky) source area in southern Slovakia, with just one piece traceable to the C2E (Mád-Erdőbénye) source area in northeast Hungary. However, four artifacts from Gorzsa that visually resemble C2E obsidian could not be matched with any known Carpathian, or indeed European, obsidian source and may derive from a previously undocumented source of obsidian or a very fine-grained obsidian-like rock.

Introduction

This chapter presents the results of portable X-ray fluorescence (pXRF) analysis of obsidian artifacts from the Late Neolithic tell settlement of Hódmezővásárhely-Gorzsa in southeast Hungary, which were recovered in excavations directed by Ferenc Horvath. The work is part of ongoing multidisciplinary research into the provenance of raw materials represented in the entire lithic assemblage from the tell excavation (Starnini et al. 2007, 2015). Considering the location of Gorzsa on the floodplain of the Tisza River and in the middle of the Alföld (the Great Hungarian Plain), the assumption is that virtually every piece of tool stone, including obsidian, was obtained from sources at least 60 km away and brought to the site either in the form of raw material or as a ready-made artifact. This presents a rare opportunity to infer the cultural connections, raw material procurement strategies, and social organization of a Late Neolithic community through geochemical identification of the sources used for the manufacture of stone tools (Szakmány et al. 2009, 2011). Choices, supply strategies, and changes in the exchange network of raw materials are the main historical issues of our scientific sourcing approach to the stone assemblage from Gorzsa.

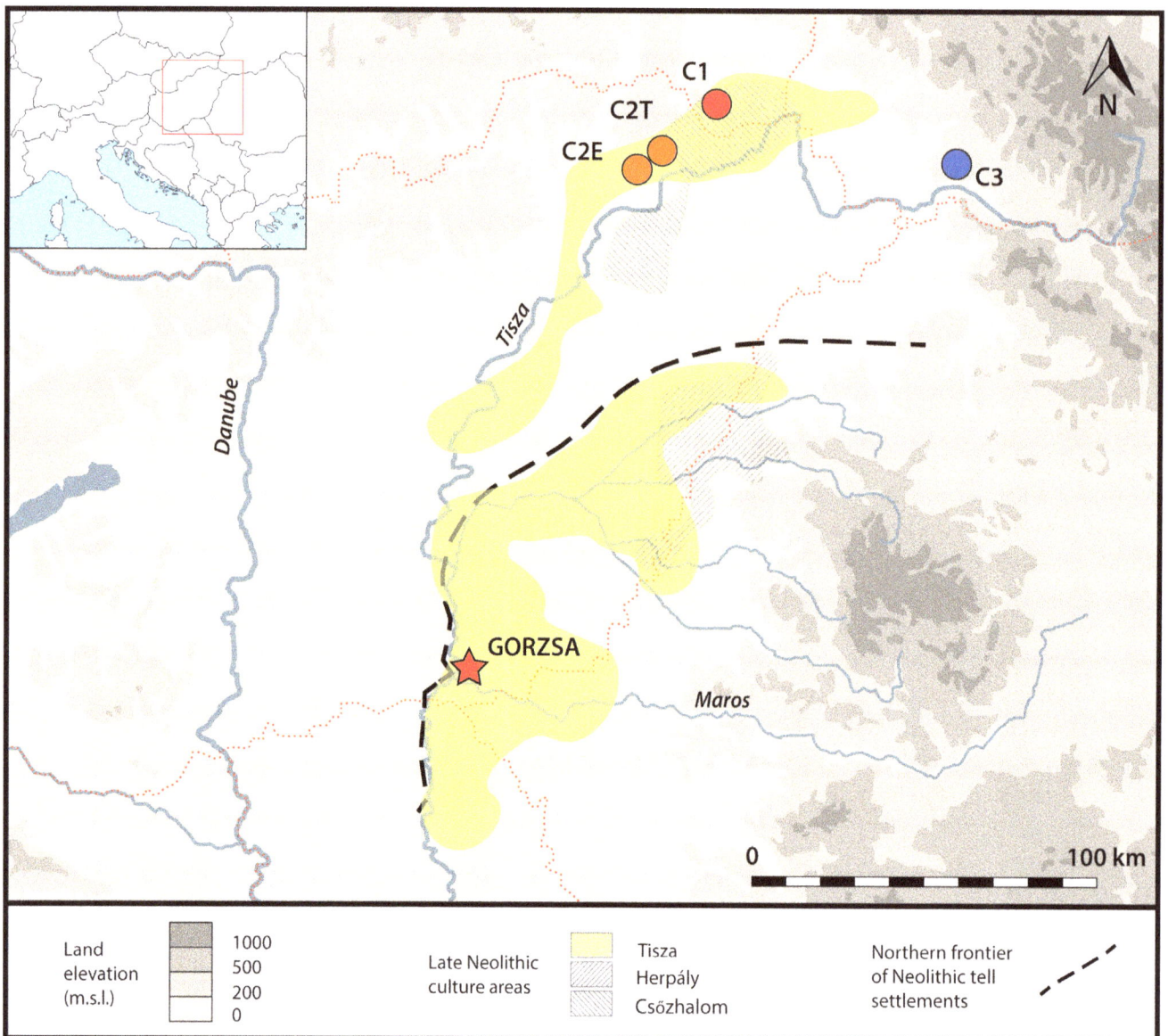

Figure 1. Map of the Alföld (Great Hungarian Plain) with the locations of Gorzsa, the Late Neolithic cultural groups (Tisza, Herpály, Csőszhalom), and the Carpathian obsidian sources (C1–C3) (drawn by C. Bonsall; cultural areas based on Raczky et al. 2020: Figure 1).

The site

The tell site of Hódmezővásárhely-Gorzsa, covering ca. five hectares, was explored during several excavation campaigns between 1978 and 1996, which investigated an area of ca. 1000 m² (Horváth 2005). The settlement was occupied during the greater part of the Late Neolithic corresponding to Phases II–V of the Tisza culture, dated to 4900–4500 cal BC, and continued to be occupied during the Bronze, Iron, and Sarmatian ages. The tell settlement is in the southern part of the Alföld (Figure 1), which during the Late Neolithic supported a dense settlement network (Tálas and Raczky 1987). The Alföld represents the northernmost frontier of the Neolithic tell settlements phenomenon that characterizes the southern Balkans. The plain is subdivided into several cultural areas: *Tisza*, located in the Tisza, Körös, and Lower Maros Valleys, *Herpály* in the Berettyó Valley, and *Csőszhalom* in the Upper Tisza Valley. The Herpály and Csőszhalom culture areas take the names of two eponymous tell sites, while Gorzsa belongs to the Tisza culture area, which takes the name of the river along which most of the sites have been found (Raczky 1992; Raczky et al. 2021). Each of these archaeological culture groups or areas corresponds traditionally to different ceramic styles, settlement types, and distribution patterns (Parkinson et al. 2018). The Alföld is delimited to the north and east by the Carpathian Mountains. In the foothills of the northeastern part of the mountain chain are located the Carpathian obsidian sources, the most important in continental Europe.

The Carpathian obsidian sources

Four chemically distinct types of obsidian are known in the region: Carpathian 1 (C1), Carpathian 2E, Carpathian 2T, and Carpathian 3. Usually, C1 and C2 can be distinguished visually: typically, C1 is highly translucent, with a glossy luster and (in some samples) darker stripes, while C2 obsidian is characteristically black or dark gray, with a duller luster, and only slightly translucent at the edges except in very thin pieces. Since the 1970s, compositional studies of the Carpathian obsidian sources have made it possible to differentiate those in northeast Hungary (C2E, C2T), eastern Slovakia (C1), and westernmost Ukraine (C3).

The C1 source area is in the Zemplín Hills of Slovakia, while the C2 source area, encompassing subgroups C2E and C2T (formerly C2b and C2a), lies in the Tokaj region of Hungary (Kaminská 2021; Furholt 2024). This division was confirmed by analyses (Biró 1984) leading to a subdivision of group C1 into subgroups C1a and C1b. Subgroup C1a comprises finds from a secondary source between Brehov and Cejkov, as well as archaeological sites ("quasi-sources") between Cejkov and Kašov, while subgroup C1b corresponds to a primary source in the Viničky-Malá Bara area (Biró and Kasztovszky 2013; Kasztovszky et al. 2014; Kaminská 2021). However, not all researchers accept the proposed subdivision of C1 obsidian (see Kohút et al. 2021). The obsidian from Transcarpathia in westernmost Ukraine was shown to be chemically distinct from the Slovakian and Hungarian sources (Rosania et al. 2008) and was designated as Carpathian 3 (C3) following the nomenclature of Williams-Thorpe and colleagues (1984). C3 obsidian is black and glassy, with macroscopic mineral grains, conchoidal to slightly hackly fracture, and is non-transparent even in thin flakes (Rácz 2018).

A general observation that emerges from studies of prehistoric sites in the Carpathian Basin is that obsidian was a lithic raw material used to produce a wide range of artifacts (endscrapers, burins, retouched blades, flakes, cores, etc.), and no close association between obsidian and a particular artifact type has been observed in any Neolithic site or culture (Kaminská 2021: 244; Starnini 1994: 57).

Figure 2. Hódmezővásárhely-Gorzsa, C1 obsidian artifacts: 1) short end scraper, used for scraping medium material (inv. MFM n. 99.3.2124./sample GOR#2124); 2) end scraper and truncation, used for scraping hard material (inv. MFM n. 99.3.2152./sample GOR#2152); 3) irregular, corticated bladelet, from the fill of Grave 4 (inv. MFM n. 99.3.2086./sample GOR#2086); 4) end scraper, used to cut soft material (inv. MFM n. 99.3.2145./sample GOR#2145); 5) micro-bladelet (inv. MFM 99.3.1720./sample GOR#1720); 6) microcore and refitting bladelet, inv. MFM 99.3.959-99.3.960./GOR#061); 7) core on a small, corticated volcanic bomb, from the ruins of House 2 (inv. MFM n. 99.3.1975./sample GOR#1975); 8) retouched blade, used for cutting medium material (inv. MFM n. 99.3.1929./sample GOR#1929); 9) corticated flake, from House 3, room 3 (inv. MFM n. 99.3.1976./sample GOR#1976); 10) mesial fragment of an unretouched blade, possibly obtained by pressure technique, used for cutting medium material (inv. MFM n. 99.3.2212./sample GOR#2122); 11) corticated and truncated bladelet, from the ruins of House 2 (inv. MFM n. 99.3.1944./sample GOR#1944) (photographs by E. Starnini).

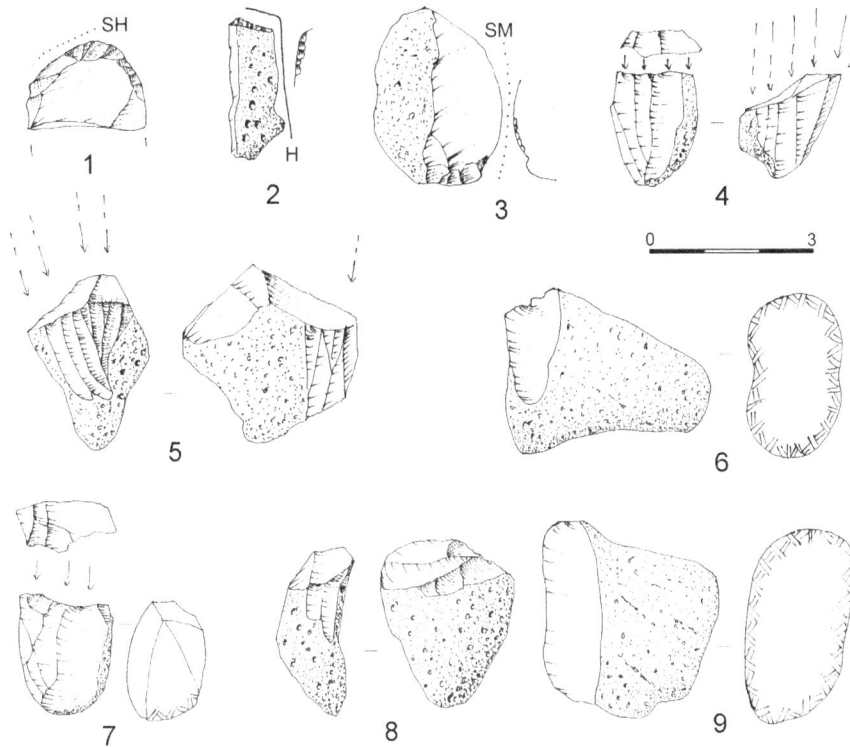

Figure 3. Hódmezővásárhely-Gorzsa, C1 obsidian artifacts: 1) short end scraper, used for scraping hard material (inv. MFM n. 99.3.431./analysis GOR#0431); 2) corticated bladelet with distal truncation, hafted (inv. MFM n. 99.3.404./analysis GOR#0404.); 3) corticated flake used for scraping soft material (inv. MFM n. 99.3.349./analysis GOR#349); 4) exhausted bladelet microcore, unidirectional, on small nodule (inv. MFM n. 99.3.341./analysis GOR#341); 5) small, exhausted bladelet core (inv. MFM n. 99.3.214.); 6) small, tested obsidian nodule (inv. MFM n. 99.3. 243.); 7) exhausted bladelet core on a small, corticated volcanic bomb (inv. MFM n. 99.3.587.); 8) pre-core on small, corticated nodule (inv. MFM n. 99.3.589.); 9) small, tested obsidian nodule (inv. MFM n. 99.3.242.) (drawings by E. Starnini).

Materials and methods

A total of 846 artifacts of obsidian were identified in excavations at Gorzsa from stratigraphic units of the Late Neolithic Tisza culture, representing ca. 20% of the knapped stone assemblages from those contexts. The obsidian assemblage comprises blanks, retouched and used artifacts, cores, and debitage waste (Figures 2 and 3). To investigate the provenance of the raw material, we carried out multielement X-ray fluorescence (XRF) analyses of a representative archaeological sample using portable/handheld instrumentation (pXRF), which is a rapid, non-destructive, and relatively inexpensive means of analyzing the chemical composition of a wide range of archaeological materials. Numerous provenance studies of obsidian have highlighted the advantages of pXRF—notably its reliability and the capability to analyze large numbers of specimens in a relatively short time (e.g., Frahm 2014; Speakman and Shackley 2013; Tykot 2018; Vazquez et al. 2012). More importantly, handheld instruments allow archaeological objects to be analyzed *in situ*, thus avoiding the problems inherent in transporting cultural heritage items from museums to laboratory facilities sometimes located abroad, as well as bypassing legal considerations and financial issues.

The XRF analyses reported here were conducted during two visits to the Móra Ferenc Múzeum in Szeged, Hungary. The first took place on August 26–28, 2019, when the measurements were taken using a *Niton XL3t Ultra He* handheld analyzer made by Thermo Fisher Scientific. Because of the COVID-19 pandemic, the second visit to Szeged had to be delayed until June 6–12, 2022. On that occasion, the instrument used was an Olympus *Vanta M* handheld XRF analyzer (Figure 4).

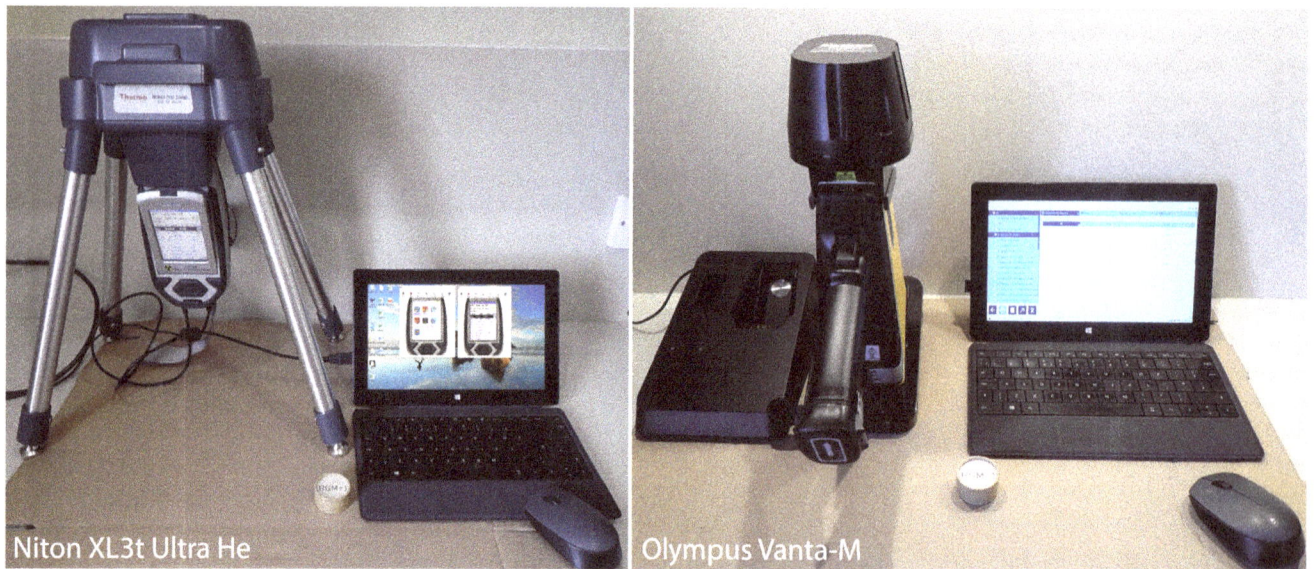

Figure 4. The pXRF analyzers used for chemical fingerprinting and source characterization of the Late Neolithic obsidian assemblage from Hódmezővásárhely-Gorzsa (photographs by C. Bonsall).

The Niton XL3t Ultra is equipped with a 2W Ag anode, 50 kV X-ray tube, and 45 mm^2 Silicon Drift Detector (SDD), while the Vanta M has a 4W Rh anode, 50 kV X-ray tube, and a 40 mm^2 SDD. Both instruments use beam filters to improve the detection of particular elements (Table 1).

Table 1. Niton XL3t Ultra and Olympus Vanta M: settings, filters, and element ranges.

Mode	kV	µA	Filter	Elements (optimized)
a) Niton XL3t Ultra: "Mining"	50	40	Main	Mn, Fe, Co, Ni, Cu, Zn, Ga, As, Se, Rb, Sr, Zr, Nb, Mo, Pd, Ag, Cd, Sn, Sb, Hf, Ta, W, Re, Au, Pb, Bi
	50	40	High	Y, Pd, Ag, Cd, Sn, Sb, Ba
	20	100	Low	Ti, V, Cr
	8	200	Light	Mg, Al, Si, P, S, Cl
b) Olympus Vanta M: "Geochem-3-beam"	40	55	Beam 1	Ti, V, Cr, Mn, Fe, Co, Ni, Cu, Zn, As, Se, Rb, Sr, Y, Zr, Nb, Mo, Ag, Cd, Sn, Sb, Ba, W, Hg, Pb, Bi, Th, U
	10	66	Beam 2	Mg, Al, Si, P, S, K, Ca, Ti, Mn
	50	65	Beam 3	Ag, Cd, Sb, Ba, La, Ce, Pr, Nd

The XL3t was configured for measuring up to 42 elements simultaneously from Mg to U in the periodic table. The analyzer was controlled from a Windows 10 laptop and operated using the factory-set Fundamental Parameter (FP) Mining Mode Calibration, using 3 of the 4 filters available (Table 1a). Each obsidian sample was measured for a total of 180s (60s per filter). The light range filter was not used, partly because the low Z elements (Al, P, Si, Cl, S, Mg) usually are not critical for obsidian characterization, and partly to reduce the overall measurement time per sample. Deploying the light range filter would have added 120s to the measurement time per sample, and ultra-low light element detection with the XL3t Ultra requires helium purge.

Measurements with the Vanta M were taken using the GeoChem 3-beam (FP calibration) model (Table 1b). The main practical advantages of the more powerful Vanta M over the XL3t Ultra are shorter measurement times and lower limits of detection (LOD). The Vanta M is also equipped with a 0.9μm-thick graphene detector window, enabling better light element detection without helium or vacuum assistance (for an assessment of the capabilities of the Vanta M, see Frahm 2017).

XRF analysis of the Gorzsa material was preceded by visual sourcing of the obsidian artifacts by Barbara Voytek and Elisabetta Starnini. The overwhelming majority were identified provisionally as C1 obsidian, with just 5 pieces attributed to a different obsidian type, assumed to be C2. Altogether, pXRF measurements were taken on 175 (ca. 21%) of the obsidian artifacts recovered from Late Neolithic contexts at Gorzsa, including all 5 pieces that had been identified provisionally as "C2 obsidian."

Results and discussion

Knapped stone artifacts can pose challenges to obtaining reliable results with non-destructive XRF analysis. They vary in size, thickness, and surface irregularity, and surfaces may be contaminated by soil or calcareous residues—all of which can affect the accuracy of XRF measurements. In addition, museum specimens often have ink or paper labels. The ideal is to clean artifacts before XRF analysis, but very often this is not possible or practical, and none of the artifacts from Gorzsa in the Móra Ferenc Múzeum could be cleaned before XRF measurements were taken. The 130 obsidian artifacts measured in 2019 were relatively large, thick pieces. However, the 45 pieces measured in 2022 included some small, thin flakes and bladelets that were narrower than the detector window of the analyzer and/or may have been of less than infinite thickness, which can also affect the accuracy of the XRF measurements. Another potential source of error was the presence of museum inventory numbers written (usually) in white ink on the flatter, ventral surfaces of flakes and blades, such that XRF measurements had to be taken on the more irregular (sometimes partially corticated) dorsal surface.

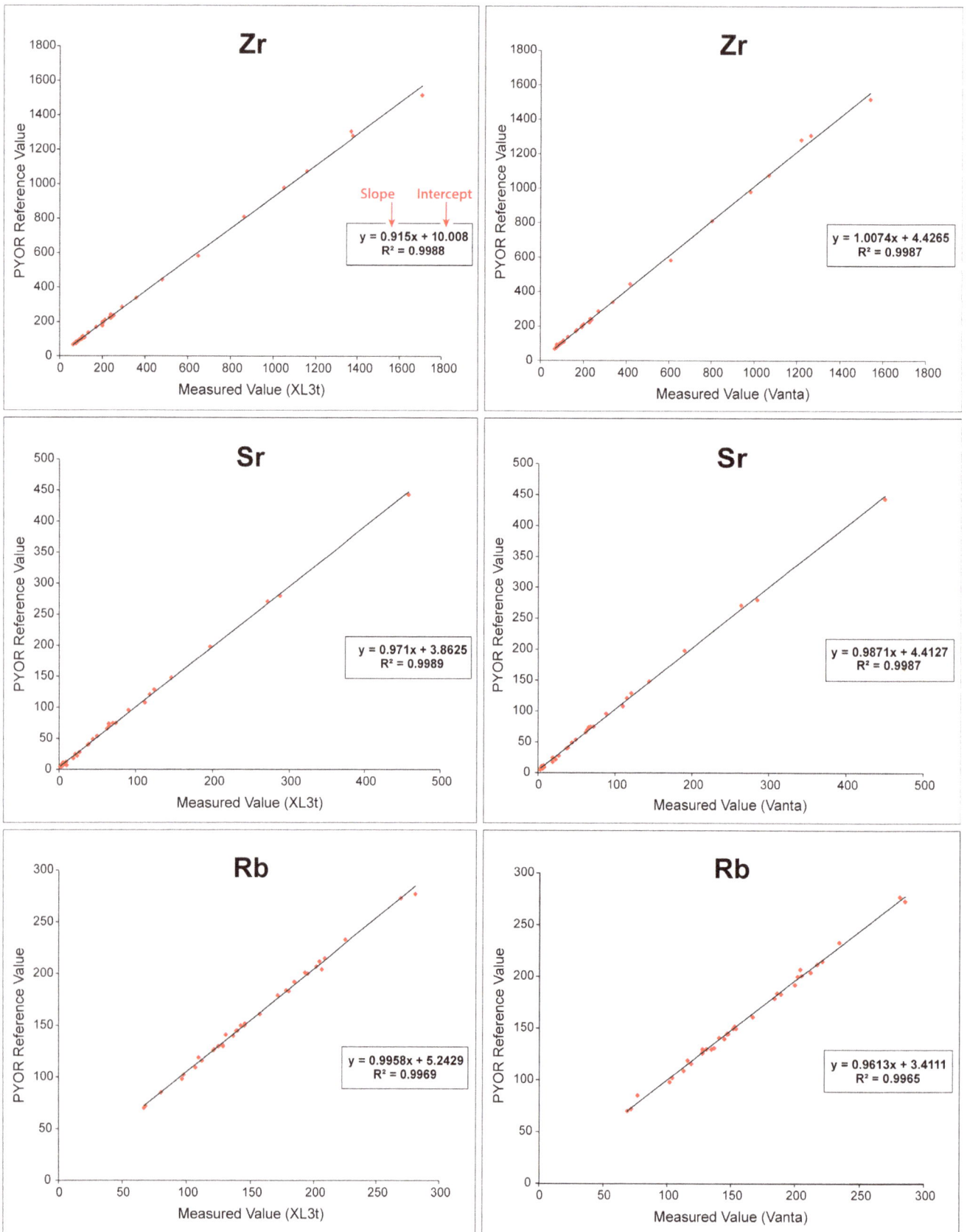

Figure 5. Calibration plots for Zr, Sr, and Rb using the Niton XL3t Ultra (left) and Olympus Vanta M (right) produced with the linear regression function (LINEST) in Excel. The equation shows the slope and intercept for the trend line. These are the calibration factors used to adjust for bias in each instrument's FP calibration model. R^2 is a measure of the strength of correlation between the measured and reference values, and ranges between 0 and 1. The closer R^2 is to 1, the stronger the relationship is (drawings by C. Bonsall).

The concentration data for Mn, Fe, Zn, Rb, Sr, Y, Zr, and Nb produced by the two pXRF analyzers were calibrated with measurements taken on the PYRO Calibration Set (Frahm 2019) using a simple linear regression model (Figure 5). Ternary plots of the Zr, Sr, and Rb values or ratios are a useful first step in assigning archaeological samples to obsidian sources. In Figure 6, the data for the Gorzsa samples are plotted against the range of variation recorded in geological specimens from the three known Carpathian source areas (C1, C2, and C3), represented by ellipsoid hulls.

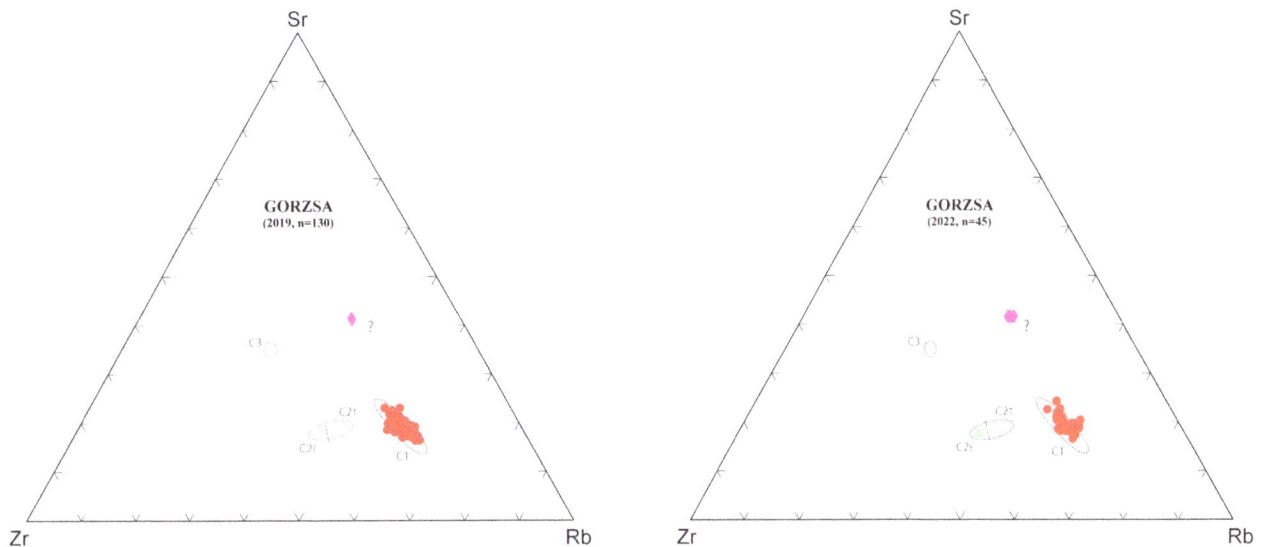

Figure 6. Ternary plots of the Zr/Sr/Rb composition of obsidian artifacts from Hódmezővásárhely-Gorzsa (symbols), and geological obsidian samples from Carpathian sources (shown as ellipsoid hulls). Red dots plot values for C1 obsidian, green triangle—C2E obsidian, and purple diamonds—unidentified "obsidian" (drawing by C. Bonsall).

Table 2. Element concentrations in parts per million in obsidian artifacts from Gorzsa (Hungary) measured by XRF using an Olympus Vanta M analyzer
(n.m. = no measurement recorded).

Sample ID	Mn	Fe	Zn	Rb	Sr	Y	Zr	Nb	Source
GOR #059	413	7172	31	195	64	32	70	10	C1
GOR #1560	404	6789	36	192	64	33	67	9	C1
GOR #1927	419	7127	31	200	62	32	70	7	C1
GOR #1928	424	7417	34	205	62	32	73	9	C1
GOR #1929	414	7218	30	189	69	31	72	7	C1
GOR #1944	401	9247	59	186	72	32	69	9	C1
GOR #1953	408	7233	43	189	61	29	61	10	C1
GOR #1958	385	7040	236	194	65	33	72	9	C1

Table 2. Continued.

Sample ID	Mn	Fe	Zn	Rb	Sr	Y	Zr	Nb	Source
GOR #1959	428	7570	33	208	67	33	69	9	C1
GOR #1961	439	8066	64	203	81	32	72	10	C1
GOR #1962	408	7060	30	202	64	32	69	9	C1
GOR #1963	438	7510	162	209	68	32	67	10	C1
GOR #1964	416	7367	47	199	63	31	77	7	C1
GOR #1965	413	6906	37	199	59	32	68	9	C1
GOR #1966	465	8735	60	206	68	28	59	n.m.	C1
GOR #1967	416	7328	61	184	59	26	57	7	C1
GOR #1968	416	7582	81	153	46	19	46	n.m.	C1
GOR #1970	385	7603	38	198	66	34	71	9	C1
GOR #1975	363	6941	43	186	68	30	66	10	C1
GOR #1976	385	6889	27	184	67	32	69	9	C1
GOR #2086	398	8375	36	189	78	32	78	10	C1
GOR #2095	417	8135	32	199	71	32	76	9	C1
GOR #2116	417	7099	31	194	63	32	70	9	C1
GOR #2124	400	6747	30	196	60	33	74	9	C1
GOR #2137	401	6630	28	189	60	32	67	9	C1
GOR #2145	430	7279	31	200	65	33	70	9	C1
GOR #2152	393	6839	31	196	62	33	75	10	C1
GOR #2162	447	7004	32	211	56	36	71	10	C1
GOR #2212	410	7007	28	193	62	32	70	10	C1
GOR #2235	401	7264	34	185	69	31	69	7	C1
GOR #99.3.14	400	6769	28	189	61	32	67	9	C1
GOR #99.3.14	428	7292	32	198	64	32	67	10	C1
GOR #x007	384	6716	30	190	61	32	67	9	C1
GOR #x007.1	388	7186	37	189	71	32	71	9	C1

Table 2. Continued.

Sample ID	Mn	Fe	Zn	Rb	Sr	Y	Zr	Nb	Source
GOR #x008	488	7571	35	200	67	34	71	10	C1
GOR #x008	488	7571	35	200	67	34	71	10	C1
GOR #x008.1	508	8521	70	199	63	28	59	n.m.	C1
GOR #x1011	396	6653	29	197	57	33	65	9	C1
GOR #x2324	402	6599	26	198	55	32	64	9	C1
GOR #2028	274	11979	41	206	81	32	175	11	C2E
GOR #x069	320	10622	76	130	139	7	66	19	?
GOR #x001	289	10986	79	131	139	6	64	17	?
GOR #337.g	302	11209	79	131	143	6	70	17	?
GOR #294.g	318	11066	80	132	143	6	69	19	?

Table 3. Element concentrations in parts per million (expressed as ranges) in geological samples of obsidian from sources in Slovakia, Hungary, and Ukraine measured by XRF using an Olympus Vanta M analyzer.

Sample locations	Mn	Fe	Zn	Rb	Sr	Y	Zr	Nb	Source
SK – Brehov-Cejkov-Kašov (n = 34)	362–466	6594–9475	27–43	180–203	56–69	31–34	62–75	7–10	C1a
SK – Viničky (n = 17)	332–408	7160–9772	27–32	166–188	69–85	29–34	70–78	9–11	C1b
HU – Mád-Erdőbénye (n = 33)	254–371	11525–17322	36–50	198–217	81–87	32–44	172–185	10–14	C2E
HU – Tolcsva (n = 11)	278–356	9712–11159	43–53	191–207	72–79	32–33	125–147	11–14	C2T
UKR – Rokosovo (n = 9)	502–594	18257–20614	55–60	148–157	181–202	23–25	209–218	11–13	C3

The Zr/Sr/Rb data for the overwhelming majority of the obsidian artifacts from Late Neolithic contexts at Gorzsa were within or close to the range determined for C1 obsidian from eastern Slovakia, confirming Voytek and Starnini's visual identifications (Tables 2–3). One piece (GOR #0028) came from the C2E source area in northeast Hungary (Figure 7, no. 5). The other four samples (GOR #x001, GOR #x069, GOR #294g, and GOR #337g; Figure 7, nos. 1–4) form a cluster that is chemically different from any known Carpathian obsidians (Figure 6).

Figure 7. Hódmezővásárhely-Gorzsa, artifacts that were visually identified as non-C1 obsidian: 1) core platform rejuvenation corniche, from square XVIII, level 7–8 (analysis GOR#x001); 2) unretouched flake fragment, from square XVIII, Pit 337 (analysis GOR#337g); 3) proximal fragment of a blade, from square III/c, level 3–4 (inv. MFM n. 99.3.2118./analysis GOR#x069); 4) unretouched flake, from square XVIII Pit 294 (analysis GOR#294g); 5) fragment of a decortication flake, C2 obsidian, from Square XVIII, level 23–24 (analysis GOR#0028) (photographs by E. Starnini).

An earlier pXRF-based characterization study by Danielle Riebe presented the compositional analysis of 203 obsidian artifacts from 7 Late Neolithic sites on the Alföld, showing that obsidian from 3 geological source areas was utilized, namely C1, C2E, and C2T (Riebe 2019). Riebe noted that artifacts made of C2E obsidian were recovered only from Tisza culture sites and that the C2T artifacts were found only at Herpály culture sites. This led her to suggest that the exploitation of the secondary obsidian sources was linked to limited access and/or sociocultural preferences.

Therefore, the analysis of obsidian artifacts from Gorzsa, another Late Neolithic site, serves also to test the hypothesis of the possible sociocultural implications of obsidian source diversification in the Alföld. Of the 203 artifacts analyzed by Riebe (2019), there were only 4 pieces of C2 obsidian (<2%). Usually, C2 obsidian is scarce after the initial stages of the Early Neolithic. Carpathian obsidians (C1, C2T, and C2E) circulated widely among the earliest farming communities of Southeast Europe, and the archaeological distribution of Carpathian obsidian coincides more or less with the territorial range of the First Temperate Neolithic, or FTN (*sensu* Nandris 1970). Hence, it serves as an important marker of the FTN interaction sphere. After ca. 5800 cal BC, nearly all obsidian found at FTN sites originated from the C1 source area (e.g., Biagi et al. 2007a, 2007b; Bonsall et al. 2017; Boroneanţ et al. 2018, 2019; Glascock et al. 2015)—a pattern that persisted into the Late Neolithic, Chalcolithic, and Early Bronze Age.

Our findings from Gorzsa are consistent with those of Riebe (2019) in documenting the presence of C2E obsidian in a Tisza culture context. However, Riebe did not consider the internal chronology of the sites she investigated. At Gorzsa, the single piece of C2E obsidian came from the earliest phase of the tell settlement, and no C2 obsidian was found in the later horizons.

The number of pieces of C2 obsidian from Late Neolithic sites on the Alföld (including its continuation into western Romania, northern Serbia, and eastern Croatia) is small, and it should not be excluded that occasional nodules of obsidian were collected from Pleistocene fluvial gravels on the plain itself. However, no pieces of obsidian in the Gorzsa collection show pebble cortex consistent with water rolling. Therefore, it is likely that all the (C1 and C2) obsidian found at Gorzsa originated from the volcanic sources on the northeastern margin of the plain. Here, it is worth emphasizing that the Late Neolithic inhabitants of Gorzsa also imported significant quantities of limnic silicite, the primary source of which occurs in the same Tokaj volcanic region containing the C2E and C2T obsidian outcrops (cf. Figure 1).

The four "obsidian" artifacts from Gorzsa that are chemically different from those made of C1 and C2E obsidian did not come from closed contexts, although at least three of them (GOR#001, GOR#294g, GOR#337g) are thought to be from the earlier (D2) and later (A–B) phases of the Late Neolithic settlement. Macroscopically, these pieces resemble obsidian: the material is black, translucent, with a glassy luster, signs of flow banding, and SiO_2 within the obsidian range (based on uncalibrated pXRF data). Chemically, this material is distinct from C1, C2E, C2T, and C3 obsidians (and the so-called C4 and C5 varieties of Rosania et al. 2008). A comparison with published compositional data for Central Mediterranean (Acquafredda et al. 2018), Aegean (Acquafredda et al. 2018; Carter et al. 2016), Central and Eastern Anatolian (Kobayashi and Mochizuki 2007), and Transcaucasian (Biagi and Gratuze 2016; Blackman et al. 1998) obsidians has so far failed to reveal a close match for the Gorzsa samples.

The "unknown" black volcanic rock from Gorzsa may be a previously undocumented variety of Carpathian obsidian or a very fine-grained volcanic rock with otherwise similar characteristics. Regardless, a Carpathian origin is supported by the fact that other lithic raw materials found at Gorzsa point to long-distance connections with the Alps, Bohemia, southern Poland, northwest Ukraine, and Bulgaria, rather than to regions with known obsidian occurrences (Starnini et al. 2015: Figure 18; Bendő et al. 2019: Figure 9).

Conclusions

Results from techno-typological analysis carried out on the obsidian assemblages from Gorzsa show a variety of products, with high percentages of categories linked to the early phases of the reduction sequence of the raw material (corticated nodules, decortication flakes, partially corticated blanks) and core exploitation and maintenance (debitage wastes, discarded irregular blanks, exhausted microcores), testifying to local reduction of the raw material and transformation of obsidian blanks.

Blades and bladelets are the most common artifact type produced with obsidian. The debitage technique employed is indirect percussion with an organic punch. Since the entire reduction sequence for obsidian seems to have occurred locally, Gorzsa most probably represents a site of reception of rough raw material pieces (i.e., nodules). Obsidian likely reached the tell in unprepared form, most probably through exchange/procurement networks connecting the southern part of the Alföld to the northern range of the Eastern Carpathians, crossing different cultural areas (Raczky et al. 2021: Figure 1).

Our pXRF analyses of the obsidian found at Gorzsa support the findings of previous researchers in showing that most obsidian consumed by Late Neolithic communities of the Alföld originated in the C1 source area in eastern Slovakia. Further, our analyses add weight to Riebe's (2019) hypothesis that the small amount of C2 obsidian that reached the Tisza culture settlements came from the C2E source in northeast Hungary.

Whether all the obsidian consumed at Gorzsa originated from Carpathian sources remains an open question. The next stage of our research will involve a more detailed study of the samples that cannot be assigned to the C1 or C2 source areas; this will allow us to determine if they are obsidian or some other black volcanic rock and will involve petrographic, SEM, and PGAA analyses. Our goal, upon completion of all analyses, will be the evaluation of patterns of change or continuity in obsidian procurement during the various phases in the lifespan of the Gorzsa tell.

Acknowledgments

This research was supported by a research scholarship at the Department of Petrology and Geochemistry, Eötvös Loránd University granted for the A.Y. 2021–2022 to one of the authors (ES) by the Tempus Public Foundation-Hungary and by the Department of Civilizations and Forms of Knowledge, Excellence Fund Project 429999_DIP_EC-CELL_CIVILTA_2018_2022 of the University of Pisa (ES).

References

Acquafredda, Pasquale, Italo M. Muntoni, and Mauro Pallara
2018   Reassessment of WD-XRF Method for Obsidian Provenance Shareable Databases. *Quaternary International* 468: 169–178.

Bendő, Zsolt, György Szakmány, Zsolt Kasztovszky, Katalin T. Biró, István Oláh, Anett Osztás, Ildikó Harsányi, and Veronika Szilágyi
2019   High Pressure Metaophiolite Polished Stone Implements Found in Hungary. *Archaeological and Anthropological Sciences* 11: 1643–1667.

Biagi, Paolo, and Bernard Gratuze
2016   New Data on Source Characterization and Exploitation of Obsidian from the Chikiani Area (Georgia). *Eurasiatica* 6: 9–36.

Biagi, Paolo, Anna Maria De Francesco, and Marco Bocci
2007a   New Data on the Archaeological Obsidian from the Middle-Late Neolithic and Chalcolithic Sites of the Banat and Transylvania. In *The Lengyel, Polgar and Related Cultures in the Middle/Late Neolithic in Central Europe*, edited by Janusz K. Kozłowski and Pál Raczky, pp. 309–326. Polish Academy of Arts and Sciences, Kraków.

Biagi, Paolo, Bernard Gratuze, and Sophie Bouchetta
2007b   New Data on the Archaeological Obsidians from the Banat and Transylvania (Romania). In *A Short Walk through the Balkans: The First Farmers of the Carpathian Basin and Adjacent Regions*, edited by Michela Spataro and Paolo Biagi, pp. 129–148. Società per la Preistoria e Protostoria della Regione Friuli, Venezia Giulia, Trieste, Italy.

Biró, Katalin T.
1984   Distribution of Obsidian from the Carpathian Sources on Central European Palaeolithic and Mesolithic Sites. *Acta Archaeologica Carpathica* 23: 5–42.

Biró, Katalin T., and Zsolt Kasztovszky
2013   Obsidian Studies Using Nuclear Techniques in Hungary. *Science for Heritage* 1: 6–9.

Blackman, M. James, Ruben Badaljan, Zaal Kikodze, and Philip Kohl
1998   Chemical Characterization of Caucasian Obsidian Geological Sources. In *L'Obsidienne au Proche et Moyen Orient : Du Volcan à l'Outil*, edited by Marie-Claire Cauvin, Alain Gourgaud, Bernard Gratuze, Nicolas Arnaud, Gérard Poupeau, Jean-Louis Poidevin, and Christine Chataigner, pp. 205–231. BAR, Oxford.

Bonsall, Clive, Nedko Elenski, Georgi Ganecovski, Maria Gurova, Georgi Ivanov, Vladimir Slavchev, V., and Radka Zlateva-Uzanova

    2017   Investigating the Provenance of Obsidian from Neolithic and Chalcolithic Sites in Bulgaria. *Antiquity* 91(356): e3, doi:10.15184/aqy.2017.

Boroneant, Adina, Cristian Virag, Ciprian Astaloş, and Clive Bonsall

    2018   Sourcing Obsidian from Prehistoric Sites in Northwest Romania. *Materiale şi Cercetări Arheologice* 14: 13–23.

Boroneant, Adina, Pavel Mirea, Anna Ilie, and Clive Bonsall

    2019   Sourcing Obsidian Artefacts from Early Neolithic Sites in South-Central Romania. *Materiale şi Cercetări Arheologice* 15: 27–40.

Carter, Tristan, Daniel A. Contreras, Kathryn Campeau, and Kyle Freund

    2016   Spherulites and Aspiring Elites: The Identification, Distribution, and Consumption of Giali Obsidian (Dodecanese, Greece). *Journal of Mediterranean Archaeology* 29(1): 3–36.

Frahm, Ellery

    2014   Characterizing Obsidian Sources with Portable XRF: Accuracy, Reproducibility, and Field Relationships in a Case Study from Armenia. *Journal of Archaeological Science* 49: 105–125.

    2017   First Hands-On Tests of an Olympus Vanta Portable XRF Analyzer to Source Armenian Obsidian Artifacts. *IAOS Bulletin* 58: 8–23.

    2019   Introducing the Peabody-Yale Reference Obsidians (PYRO) Sets: Open-source Calibration and Evaluation Standards for Quantitative X-ray Fluorescence Analysis. *Journal of Archaeological Science: Reports* 27: 101957.

Furholt, Kata

    2024   Depositional Patterning of Obsidian Artifacts: Studying Diverse Value Concepts in the Neolithic Carpathian Basin. In *Reflections on Volcanic Glass: Proceedings of the 2021 International Obsidian Conference*, edited by Lucas R. M. Johnson, Kyle P. Freund, and Nicholas Tripcevich, pp. 15–47. ARF Monographs, Berkeley, California.

Glascock, Michael D., Alex W. Barker, and Florin Draşovean

    2015   Sourcing Obsidian Artifacts from Archaeological Sites in Banat (Southwest Romania) by X-Ray Fluorescence. *Analele Banatului, S.N., Arheologie – Istorie* 23: 45–50.

Horváth, Ferenc

    2005   Gorzsa. Előzetes eredmények az újkőkori tell 1978 és 1996 közötti feltárásából [Gorzsa. Preliminary Results of the Excavation of the Neolithic Tell between 1978 and 1996]. In *Hétköznapok Vénuszai*, edited by Livia Bende and Gábor Lőrinczy, pp. 51–83. Móra Ferenc Múzeum, Szeged, Hungary.

Kaminská, Ľubomíra
2021   Use of Obsidian in Slovak Prehistory. In *Fossile Directeur, Multiple Perspectives on Lithic Studies in Central and Eastern Europe*, Študijné *zvesti Archeologického* ústavu *SAV –Supplementum* 2: 231–250.

Kasztovszky Zsolt, Katalin T. Biró, and Zoltán Kis
2014   Prompt Gamma Activation Analysis of the Nyírlugos Obsidian Core Depot Find. *Journal of Lithic Studies* 1: 151–163.

Kobayashi, Katsuji, and Akihiko Mochizuki
2007   Source Identification of Obsidian Projectile Points from Kaman-Kalehöyük. *Anatolian Archaeological Studies* 16: 177–182.

Kohút, Milan, John A. Westgate, Nicholas J. G. Pearce, and Pavel Bačo
2021   The Carpathian Obsidians – Contribution to Their FT Dating and Provenance. *Journal of Archaeological Science: Reports* 37 (June): 102861.

Nandris, John
1970   The Development and Relationships of the Earlier Greek Neolithic. *Man* 5(2): 192–213.

Parkinson, William A., Attila Gyucha, Panagiotis Karkanas, Nikos Papadopoulos, Georgia Tsartsidou, Apostolos Sarris, Paul R. Duffy, and Richard W. Yerkes
2018   A Landscape of Tells: Geophysics and Microstratigraphy at Two Neolithic Tell Sites on the Great Hungarian Plain. *Journal of Archaeological Science: Reports* 19: 903–924.

Rácz, Béla
2018   The Carpathian 3 Obsidian. *Archeometriai Műhely* XV: 181–186.

Raczky, Pál
1992   The Tisza Culture of the Great Hungarian Plain. *Studia Praehistorica* 11–12: 162–176.

Raczky, Pál, András Füzesi, Katalin Sebők, Norbert Faragó, Péter Csippán, and Alexandra Anders
2020   A Special House from the Late Neolithic Tell Settlement of Berettyóújfalu-Herpály (Hungary). Reconstruction of a Two-Storey Building, Its Furnishings and Objects from the Earlier 5th Millennium BC. *Studia Troica* 11: 429–457.

Raczky, Pál, András Füzesi, Knut Rassmann, and Eszter Bánffy
2021   Újkőkori települési halmok Hódmezővásárhely környékén. In *Korok, kultúrák, lelőhelyek*, edited by Csányi Viktor, pp. 7–26. Tornyai János Múzeum, Hódmezővásárhely, Hungary.

Riebe, Daniela J.
2019   Sourcing Obsidian from Late Neolithic Sites on the Great Hungarian
Plain: Preliminary p-XRF Compositional Results and the Socio-Cultural
Implications. *Interdisciplinaria Archaeologica* 10(2): 113–120.

Rosania, Corinne N., Matthew T. Boulanger, Katalin T. Biró, Sergey Ryzhov, Gerhard
Trnka, and Michael D. Glascock
2008   Revisiting Carpathian Obsidian. *Antiquity* 82(318), http://www.antiquity.
ac.uk/projgall/rosania318/, accessed February 15, 2022.

Speakman, Robert J., and M. Steven Shackley
2013   Silo Science and Portable XRF in Archaeology: A Response to Frahm.
Journal of Archaeological Science 40: 1435–1443.

Starnini, Elisabetta
1994   Typological and Technological Analysis of the Körös Culture Chipped,
Polished and Ground Stone Assemblages of Mèhtelek-Nàdas (N-E Hungary). *Atti
della Società per la Preistoria e Protostoria della regione Friuli Venezia-Giulia* 8: 29–96.

Starnini, Elisabetta, Barbara A. Voytek, and Ferenc Horváth
2007   Preliminary Results of the Multidisciplinary Study of the Chipped Stone
Assemblage from the Tisza Culture Site of Tell Gorzsa (Hungary). In *The Lengyel,
Polgár and Related Cultures in the Middle/Late Neolithic in Central Europe*, edited by
Janusz K. Kozłowski and Pál Raczky, pp. 269–278. The Polish Academy of Arts
and Sciences, Kraków.

Starnini, Elisabetta, György Szakmány, Sándor Józsa, Zsolt Kasztovszky, Veronika
Szilágyi, Boglárka Maróti, Barbara Voytek, and Ferenc Horváth
2015   Lithics from the Tell Site Hódmezővasárhely-Gorzsa (S-E Hungary):
Typology, Technology, Use and Raw Material Strategies During the Late
Neolithic (Tisza Culture). In *Neolithic and Copper Age between the Carpathians
and the Aegean Sea. Chronologies and Technologies from the 6th to 4th Millennium
BC. International Workshop Budapest 2012*, edited by Svend Hansen, Pál Raczky,
Alexandra Anders, and Agathe Reingruber, pp. 115–138. Habelt Verlag,
Bonn, Germany.

Szakmány, György, Elisabetta Starnini, Ferenc Horváth, Veronika Szilágyi, and
Zsolt Kasztovsky
2009   Investigating Trade and Exchange Patterns during the Late Neolithic: First
Results of the Archaeometric Analyses of the Raw Materials for the Polished and
Ground Stone Tools from Tell Gorzsa (Southeast Hungary). In *Proceedings of the
6th Meeting for the Researchers of Prehistory: Raw Materials and Trade, Kőszeg, March
19–21, 2009*, edited by Gábor Ilon, pp. 369–383. Vas County Museums Directorate,
Szombathely, Hungary.

Szakmány, György, Elisabetta Starnini, Ferenc Horváth, and Balázs Bradák
2011   Investigating Trade and Exchange Patterns in Prehistory: Preliminary
Results of the Archaeometric Analyses of the Stone Artifacts from Tell Gorzsa
(Southeast Hungary). In *Proceedings of the 37th International Symposium on
Archaeometry, Siena, May 12–16, 2008*, edited by Isabella Turbanti-Memmi, pp.
311–319. Springer, Berlin.

Tálas, László, and Pál Raczky (editors)
1987   *The Late Neolithic of the Tisza Region. A Survey of Recent Excavations and their
Findings: Hódmezővásárhely-Gorzsa, Szegvár-Tűzköves,* Öcsöd-Kováshalom, *Vésztő-
Mágor, Berettyóújfalu-Herpály.* Szolnok, Budapest.

Tykot, Robert H.
2018   Portable X-ray Fluorescence Spectrometry (pXRF). In *The Encyclopedia of
Archaeological Sciences*, edited by Sandra L. López Varela, pp. 1–5. Wiley, Malden,
Massachussetts.

Vazquez, Cristina, Oscar Palacios, Marcó Parra Lue-Meru, Graciela Custo, Martha
Ortiz, and Martín Murillo
2012   Provenance Study of Obsidian Samples by Using Portable and
Conventional X-ray Fluorescence Spectrometers. Performance Comparison
of Both Instrumentations. *Journal of Radioanalytical and Nuclear Chemistry*
292: 367–373.

Williams-Thorpe, Olwen, Stanley E. Warren, and John G. Nandris
1984   The Distribution and Provenance of Archaeological Obsidian in Central
and Eastern Europe. *Journal of Archaeological Science* 11: 183–21

CHAPTER 4

Technological Analysis and Geochemical Characterization of Obsidian Artifacts from the Middle Neolithic Site of Opatów, Southeast Poland

DAGMARA H. WERRA, MARCIN SZELIGA, AND RICHARD E. HUGHES

Dagmara H. Werra Institute of Archaeology and Ethnology Polish Academy of Sciences, 105 Solidarności Avenue, 00-140 Warsaw, Poland (dwerra@iaepan.edu.pl), ORCID 0000-0003-2233-1663

Marcin Szeliga Institute of Archaeology, Maria Curie-Skłodowska University in Lublin, 4 M. C.-Skłodowska sq, 20-031 Lublin, Poland (marcin.szeliga@mail.umcs.pl), ORCID: 0000-0002-5185-073X

Richard E. Hughes Geochemical Research Laboratory, 2440 East Tiffany Lane, Sacramento, CA 95827-1433, USA (rehughes@silcon.com)

Abstract

This chapter discusses the obsidian assemblages from Opatów, site 2, one of the most interesting archaeological sites from the Middle Neolithic in Southeast Poland. The site is the eponymous site of the Samborzec-Opatów group of the Lengyel culture (c. 4900–4700 cal BC). The results of morphological and technological characterization, as well as non-destructive energy dispersive X-ray fluorescence (EDXRF), of 264 obsidians artifacts from this site are presented. EDXRF results show that all artifacts analyzed were manufactured from volcanic glass of the Carpathian 1 chemical type, the origin of which is located in Slovakia. The results of morphological and technological analysis show that the processing of obsidian within the site was most likely concentrated only on obtaining very fine, even microlithic, blade blanks. All the specimens are physically small, so the question remains whether this is due to the natural properties of the raw material or to the cultural norms that guided the knapping technology.

Introduction

In Poland, especially in the Upland area, good quality raw materials, mainly flint, are abundant and easily procured. The Vistula and Odra River areas contain deposits of several kinds of flint, some of very high quality, which were widely used in prehistoric times. Despite this wealth in local raw materials, archaeological sites also contain so-called exotic raw materials conveyed from hundreds of kilometers beyond the border of Poland. One of these "foreign" materials is obsidian, favored for artifact manufacturing because of its unique properties, such as excellent cleavage and sharp cutting edges. The most important obsidian outcrops for prehistoric communities in Central Europe are located in the vicinity of the Zemplén Mountains in southeast Slovakia and northeast Hungary (Figure 1; Thorpe et al. 1984; Rosania et al. 2008; Přichystal 2013, 161; Rácz 2018; Werra et al. 2021).

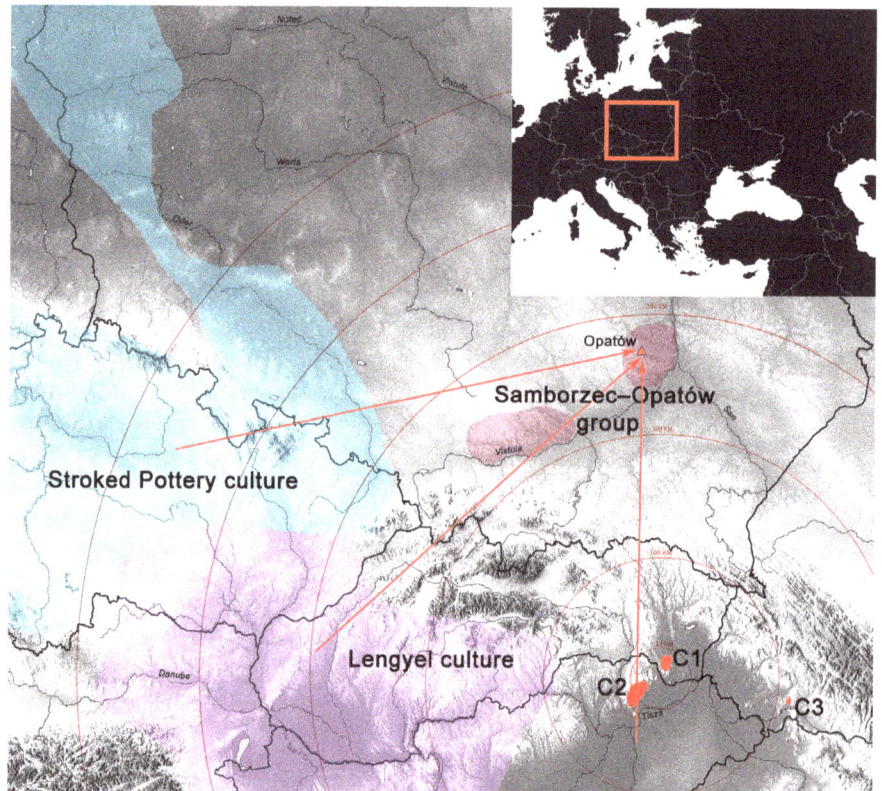

Figure 1. East-Central Europe ca. 4900–4400 calBC. C1 – Carpathian 1 geological obsidian outcrops; C2 – Carpathian 2 geological obsidian outcrops; C3 – Carpathian 3 geological obsidian outcrops. Red lines mark distances from Carpathian 1 source locations. Red arrows indicate directions of cultural influences. After Kadrow and Zakościelna 2000 and Řídký et al. 2015, with changes.

In Poland, the oldest obsidian artifacts are connected with the activity of prehistoric communities of the Paleolithic and Mesolithic periods (Hughes et al. 2018), but they are very rare. With the arrival of the first Neolithic societies (Linear Pottery Culture, c. 5500–4900 calBC), the increase of obsidian can be observed throughout the Middle and Late Neolithic (ca. 5000–4300/4200 calBC) but declined most probably in the Early Eneolithic (at least ca. 4000/3900 calBC; cf. Szeliga 2021). Because no natural outcrops of obsidian occur in Poland, the material used to make all of the obsidian artifacts present on Polish archaeological sites must have been conveyed there by some means (Szeliga 2009; Wilczyński 2016; Szeliga et al. 2021).

One of the most interesting archaeological sites from the Middle Neolithic in Poland, due to its exceptional artifact assemblages and the presence of large numbers of obsidian artifacts, is Opatów, Opatów district, in Southeast Poland. This site is representative of the Samborzec-Opatów group (ca. 4900–4700 calBC), which is known only from a few scattered sites in the loess uplands of the Upper Vistula basin (Figure 1).

The research presented here describes the obsidian knapping technology, the morphological types recorded at the Opatów site, the distribution of obsidian artifacts, and the issues of conveyance and contacts among prehistoric communities. We also touch on the role (cultural/social/symbolic) of obsidian among Middle Neolithic communities of the Samborzec-Opatów group of the Lengyel culture.

Opatów, site 2: a short history of long research

The site in Opatów is located on the Sandomierz Upland on the slope of the left bank of the Opatówka River, which is the left tributary of the Vistula (Figure 1). The archaeology of the region of Opatów was studied in 1913–1915 by Józef Milicer, an amateur archaeologist and member of the Polish Sightseeing Society (*Polskie Towarzystwo Krajoznawcze*). In the interwar period (in 1922, 1924, and 1929) excavations were carried out by Zofia Podkowińska. During the research in the interwar period, 38 features related to the activity of Neolithic (post-linear) communities and the Bronze Age Trzciniec culture were examined. Unfortunately, the report on the work that had been prepared for printing was completely destroyed in 1939 during World War II (Kulczyński and Pyzik 1959; Kosterski-Spalski 1963; Podkowińska 1968; Jędrzejczyk 2013).

Podkowińska continued her research in Opatów after the war by carrying out excavations in 1965 with Leokadia Wrotek. The research was undertaken by the Department of Neolithic and Early Bronze Age of what was then the Institute of the History of Material Culture of the Polish Academy of Sciences (IHKM PAN; now the Institute of Archaeology and Ethnology, Polish Academy of Sciences). Some articles have appeared based on the results (Podkowińska 1953, 1968). The only fuller study of the materials from Opatów, apart from reporting articles, is the study of lithic inventories compiled by Hanna Więckowska (1927–2013), an archaeologist from the IHKM PAN (Więckowska 1971; Kowalewska-Marszałek 2004, 2007, 2012).

During the investigations of the 1920s, the site attracted the attention of archaeologists due to its unique character and high proportion of obsidian products. Researchers also saw analogies in the ornamentation of ceramic vessels recovered at Opatów with the materials from the sites in Lesser Poland and the Czech Republic, and the closest parallels in the materials from Samborzec, in Sandomierz district (Podkowińska 1968).

The unevaluated character of the archaeological materials from the sites in Samborzec and Opatów prompted Podkowińska to separate the Samborzec-Opatów group of the Lengyel culture (ca. 4900–4700 calBC; see Figure 1). The group was defined by Podkowińska (1953: 32–44) and then later redefined many times (Kozłowski 1966; Kamieńska 1967; Kamieńska and Kozłowski 1970; Kozłowski and Kozłowski 1977; Kaczanowska and Kozłowski 1994; Kadrow and Zakościelna 2000; Kowalewska-Marszałek 2004; Kozłowski 2004; Kulczyc-ka-Leciejewiczowa 2004; Zápotocká 2004).

This cultural group represents the oldest horizon of the Danubian communities on the northern side of the Carpathians after the decline and disappearance of the Linear Pottery culture (abbreviated as LBK, from German; Linearbandkeramik). It appears to have developed under the strong influences of the Lengyel culture, the Stroked Pottery culture, and the sociocultural systems from the Upper Tisza River basin (Figure 1). These multidirectional cultural interactions had a crucial influence on the syncretic character of the Samborzec-Opatów group, manifested mainly in its ceramics. In the Sandomierz Upland, the pottery assemblages are distinguished by the frequent presence of vessels with bulging necks and thick walls, with horizontal bands dominating the ornamentation. The lithic industry is prevailed by "chocolate" flint with the frequent presence of obsidian items (Kaczanowska 1990; Kaczanowska and Kozłowski 1994; Kadrow and Zakościelna 2000, 196).

This unique character, in turn, has had a fundamental impact on the classifications of this cultural group within the framework of the basic cultural taxonomy. The materials of the Samborzec-Opatów group from western Lesser Poland (Southern Poland) have been classified as part of the Lengyel culture, and those from the Sandomierz Upland (including the site in Opatów) were defined as an "early Malice-Lengyel mixed group" (see references in Kadrow and Zakościelna 2000). Regardless of the recent discussions of the cultural classification of the Samborzec-Opatów group and the archaeological site in Opatów, this obsidian assemblage has played an important role in the study of obsidian conveyance to the north of the Carpathians in the Neolithic, including the continuity and intensity of this phenomenon after the disappearance of the LBK.

During the excavations in the 1920s and the 1960s at the Opatów site, almost 300 obsidian artifacts were recovered (5.04% of the complete lithic assemblage). Hitherto, this is the largest and most chronologically undisturbed obsidian assemblage related to the oldest post-linear cultural horizon in the region of the Upper Vistula basin. The technological, morphological, and chemical characteristics of this material, as well as refitting analysis, are presented below. All those analyses were undertaken to illustrate the method of obsidian knapping by Middle Neolithic communities and to thus obtain a picture of the everyday life of those prehistoric communities, as well as to analyze the presence of obsidian in the sociocultural context.

Technological characterization of the obsidian inventory, classification of products, and methodological assumptions of the conducted analyses

The obsidian products from Opatów were processed according to a classification system referring to the methodological solutions developed and most often used in previous studies on flint-making of the Danube communities on the northern side of the Carpathians (Kozłowski and Kulczycka 1961; Dzieduszycka-Machnikowa and Lech 1976; Lech 1981, 1997, 2008; Kaczanowska 1985; Kaczanowska et al. 1987; Małecka-Kukawka 1992; Zakościelna 1996). They are based on a division and analysis of finds within four basic morphological groups: I) cores; II) blades and fragments; III) flakes and chips; and IV) implements (retouched tools). These groups correspond to the basic sequence of processing activities related to the preparation and exploitation of cores, and to obtaining flake and blade blanks, and producing tools.

The analysis of products within the individual morphological groups was carried out based on research methods that accounted for the sets of both universal and strictly individual morphological and metric properties of the studied assemblage. The observed variations were analyzed using basic tools and statistical methods. In the case of the metric properties, the subject of the research was basic sizes (length, width, and thickness) and other measurable diagnostic parameters (e.g., platform sizes), as well as the weight of individual products.

The analytical data gathered allowed for the presentation of the morphological and typological characteristics of each of the groups of the assemblage, along with the determination of the degree of formal differentiation of the categories of finds included in them. The whole dataset thus constitutes a primary point of reference in an attempt to assess the extent and nature of obsidian processing by Neolithic communities on the site.

In parallel to the morphometric analyses of the finds, intensive refitting exercises were also carried out. This work was carried out on two levels, taking into account the different scope and character of the exploitation of the site (Hofman 1992: 10). First, the refitting procedure was applied to products from the fills of individual negative features, and second, potential joins between products from various features were sought. Unfortunately, the effects of this work turned out to be far from what was expected. Despite the presence of objects with very similar macroscopic properties within individual assemblages of finds (undoubtedly produced by the same production or rejuvenation activities carried out in the case of the same cores), in the course of the refitting, it was not possible to determine any certain joins between individual finds. The only exception is a refitted tip and butt of a retouched blade from Pit 40 (Figure 11: 7). At the same time, it should be emphasized that, in this case, there is no certainty that the fracture was intentional and occurred during the operation of the analyzed settlement.

The main reason for the inability to refit most of the material seems to be the limitations resulting from the specificity and nature of the assemblage from Opatów. This is in addition to the undoubtedly reduced size of the collection in relation to the initial state (originally, the inventory consisted of 292 items; Więckowska 1971). The problem is obviously much wider and concerns not only the analyzed obsidian materials but all the flint and stone assemblages discovered within Neolithic settlements. Groups of products lying within the fills of features constitute assemblages selected and modified already at the stage of creation and use (that is, by an extemporary selection of products by producers and users). Moreover, they are at least partially redeposited in relation to the place of their origin (de Grooth 1981, 117; 1990, 198) and depleted as a result of natural (e.g., intense slope erosion) and anthropogenic and cultural post-deposition processes (especially settlement activity in later stages of use of the settlement—compare, for example, the analyses of Paleolithic and Mesolithic communities in Tomaszewski 1986; Fiedorczuk 1992, 2006; Wąs 2005). This problem was raised in the case of the rich lithic collections originating from the sites of the Rhineland cluster of the LBK. In the case of the Langweiler 8 site, the total number of flint products related to the settlement of the LBK culture was about 10,000 items, accounting for only 10–15% of the original content of the inventories (de Grooth 1990: 203).

A very significant difficulty in the case of the Opatów inventory was also the very small size of the group of obsidian products and their specific, highly uniform macroscopic properties (high transparency). These circumstances undoubtedly had a decisive influence on the almost total lack of success of the refitting method used in the case of this assemblage of material.

The quantity and structure of the inventory

A total of 264 obsidian products were analyzed, constituting approximately 90.4% of the original inventory (Więckowska 1971). These artifacts were obtained primarily during

Table 1. Opatów, Site 2. Quantitative and qualitative structure of obsidian artifacts discovered within the individual features and secondarily deposited (Trench 1924, 1929).

Features	Cores	Blades	Flakes and scraps/chips	Retouched tools	Total
1	-	1	-	-	1
2	-	-	1	-	1
4	-	-	1	-	1
7	1	3	1	1	6
21	-	1	-	-	1
35	-	32	36	3	71
36	-	-	1	-	1
37	8	109	28	6	151
40	2	8	12	1	23
40A	-	-	1	-	1
Trench 1924/1929	2	3	1	1	7
Total	13	157	82	12	264

the exploration of 10 features, and only in a few cases were they secondarily deposited in the layers outside these features (Table 1).

The current structure of the inventory is characterized by the considerable domination of the blades group, with almost two times fewer flakes and a minimum frequency of cores and prepared tools (Figure 2). These relations, very clearly visible in the case of the structure of the entirety of obsidian materials, are also clear in the individual distributions within the larger assemblages from the feature fills—for example, Pits 35 and 37 (Table 1). They correspond to the structures of assemblages that are typical for the so-called utilitarian residue, where the processing of raw materials was generally carried out to a small, or even minimal, extent (e.g., Lech 2008).

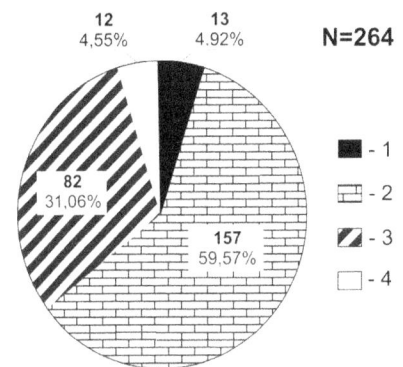

Figure 2. Opatów, Site 2. Morphological structure of the obsidian artifacts: 1 – cores; 2 – flakes and chips; 3 – blades and their fragments; 4 – retouched tools.

Morphometric and typological analysis

Cores

A total of 13 specimens were classified within this group (4.92% of the entire collection; see Figure 2). They represent in all cases partially used or fragmented blade core forms, abandoned at a more or less advanced stage of exploitation. This is manifested in most cases by their very small or even microlithic dimensions and weight (Table 2), as well as by the very small size of the scars of the blade blanks obtained from them. The length of blade negatives preserved within the flaking surface of individual cores ranges from 14.4 to 27 mm, with a clear predominance of specimens measuring 18–22 mm. The width of the final blades detached from the core is mostly within the range of 3–10 mm (Figure 3).

Table 2. Opatów, Site 2. Dimensions (mm) and weight (g) of the analyzed cores.

Features	Inventory number	Height	Width	Thickness	Weight	Figures
7	#002	26.5	29	25	21.83	Figure 6: 7
37	#086	18.5	15,5	23,5	5.88	Figure 6: 2
	#087	29	28	20	14.19	Figure 5: 5
	#088	22.5	17	22	6.60	Figure 5: 6
	#089	26	28	26	17.17	Figure 5: 4
	#090	19.5	23	21,5	7.65	Figure 5: 7
	#091	29	18	23	10.09	Figure 5: 3
	#092	29	20	26	11.65	Figure 6: 1
	#093	23	21	19	8.82	Figure 5: 8
40	#111	17.5	14	14	3.99	Figure 6: 3
	#121	18.5	27.5	23	12.41	Figure 6: 5
Trench 1924/1929	#256	21	22	15	7.38	Figure 5: 1
	#257	17.5	12	20.5	4.09	Figure 5: 2

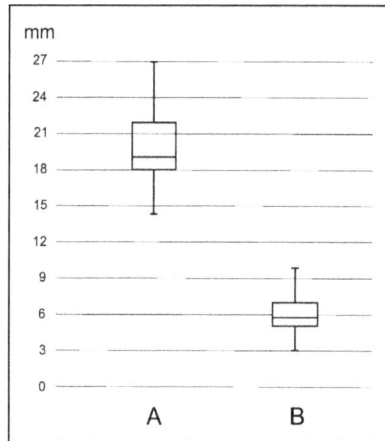

Figure 3. Opatów, Site 2. Dimensions of blade scars preserved on the cores: A – length; B – width.

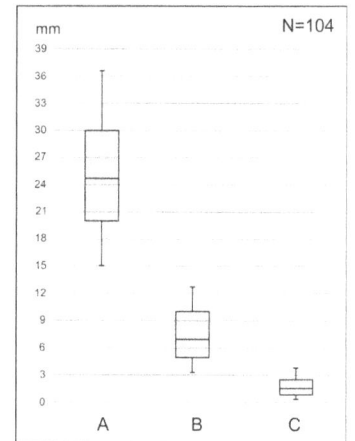

Figure 4: Opatów, Site 2. Dimensions of complete blades: A – length; B – width; C – thickness.

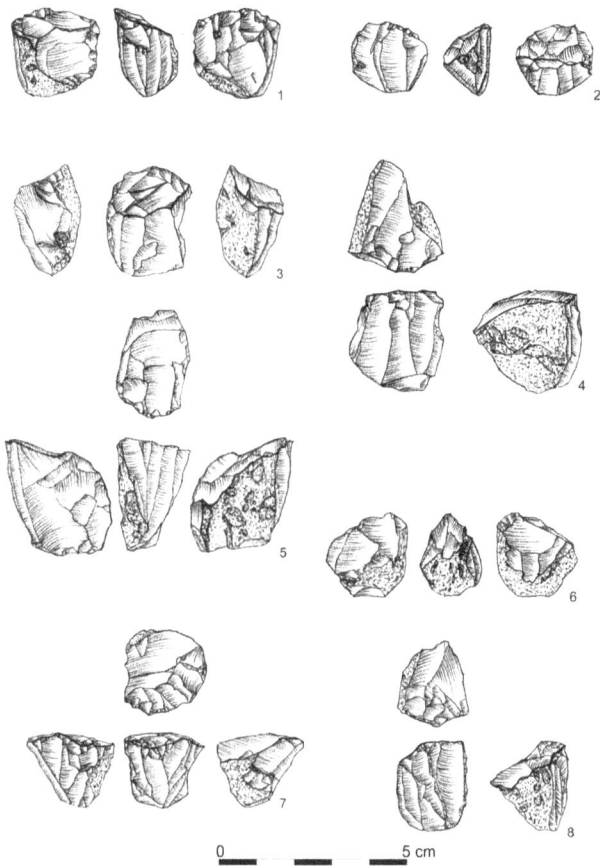

Figure 5: Opatów, Site 2. Cores obtained from Trench 1924/1929 (1–2) and from Feature 37 (3–8).

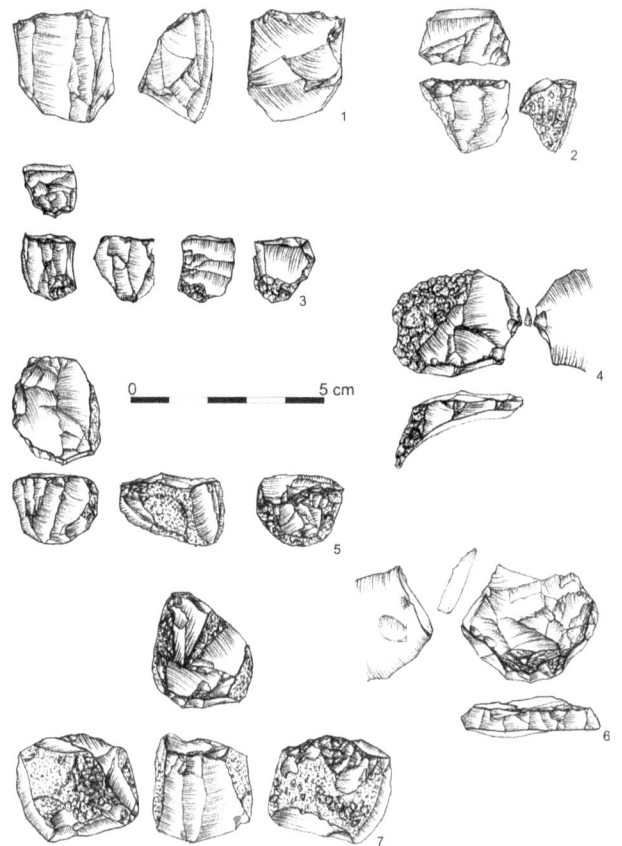

Figure 6: Opatów, Site 2. Cores (1–3, 5, 7–8) and rejuvenation flakes (4, 6) from Features: 37 (1–2), 40 (3–5), 35 (6) and 7 (7).

These data correspond closely with the results of measurements of the analogous metric properties of the whole blades recovered (Figure 4).

All cores recorded were single-platform blade cores: in typological terms, mainly the sub-conical or conical type (Figures 5: 1–3, 7–8; 6: 1–3). Some were of sub-keeled (Figures 5: 4–5; 6: 5) or polyhedral form (Figure 6: 7). Their striking platforms were formed with the detachment of several flakes, usually struck from the flaked surfaces (e.g., Figure 5: 1). Sometimes, however, they were flaked also from one or both sides of the core (e.g., Figures 5: 2; 6: 5, 7). In three specimens, the surface of the previous primary flaking surface was used as the platform. In these cases, it was associated with a perpendicular change of the core orientation. In two cases, the new flaking surface was located within the existing platform (Figures 5: 4–5). In a single case, it was in the posterior of the core (Figure 5: 1). In the case of two cores (Figures 5: 2, 6), the new flaking surface was located within the existing rear, while the striking platforms remained unchanged in the same place.

The treatments related to the preparation of the cores identified in the analyzed material are mainly limited to the preparation of the platform. Only sporadically were traces of removal of cortical flakes found on the back, top, or one of the sides of the core (e.g., Figures 5: 3; 6: 1). The cortex was most often left on the sides of the core. In one case, a fragment of a side-crest has been preserved (Figure 5: 5). The occasional use of rejuvena-

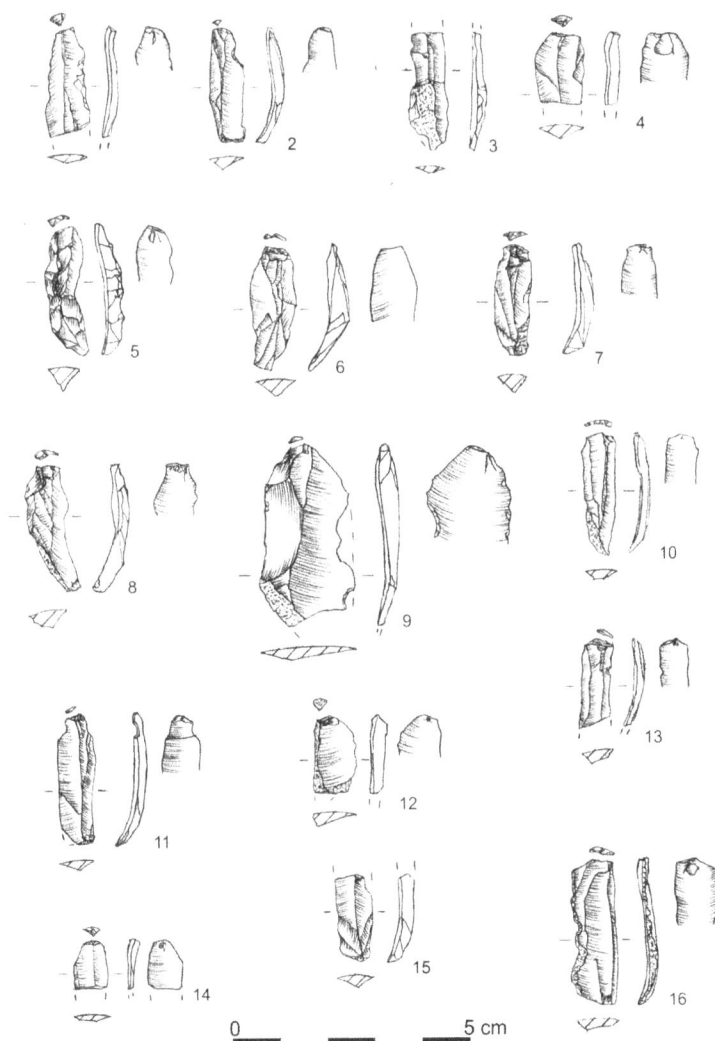

Figure 7: Opatów, Site 2. Selection of blades from Features: 37 (1–8, 10, 13), 1 (9), 7 (11, 14–16) and 21 (12).

tion treatments on obsidian nodules is also confirmed by the negligible quantity of blades showing different parts of the preparation of the crests in the assemblage (Figure 7: 5–6).

Various types of rejuvenation procedures were used slightly more often during the exploitation of the cores. In addition to the aforementioned change of orientation, they mainly consisted of the correction of the core angle, performed on an *ad hoc* basis and immediately before the knapping of subsequent blades. These operations involved, on the one hand, quite a delicate rejuvenation of the striking platforms of the cores, carried out mainly from the side of the flaking surface, and to a lesser extent, from the sides of the core. In addition, regular trimming of the platform edges took place. The traces of this operation are single or multiple small flake scars irregularly spaced along the edges of the striking platforms of cores (e.g., Figures 5: 1–8; 6: 1–2), as well as the presence in the assemblage of small rejuvenation flakes (Figure 5: 2). The striking platforms of the cores were sometimes completely renewed. This was carried out by removing their entire surface, or the larger part of them, with single blows from the side of the flaking surfaces or one of the sides. The remnants of these treatments are a few rejuvenation core tablets in the assemblage (e.g., Figures 6: 4, 6). The high frequency of trimming of the platform edges is also shown by the high frequency of the characteristic remains registered in the dorsal surfaces of the proximal parts of the analyzed blades, and by the prevalence of faceted butts seen among them.

Blades and their fragments

Blades and their fragments constitute the most numerous category of items in the analyzed collection, represented by a total of 157 specimens (59.47% of the total; see Figure 2). The analysis of the state of preservation shows a clear preponderance of fragments over those preserved in their entirety (the overall percentage of the latter was 31.21%). Among the incomplete specimens, fragments containing the butt of the blade were clearly dominant, significantly exceeding the number of fragments of the middle and distal parts (Table 3). The observed proportions show very close analogies to the state of preservation of blades recorded for many other Neolithic sites (e.g., Lech 1981; Kaczanowska et al. 1987; Kozłowski 1989; Małecka-Kukawka 1992; Mateiciucová 2008). This suggests at least partially intentional breaking of the blades in terms of correcting the slenderness and the degree of curvature of individual specimens (e.g., Kaczanowska 1971; Dzieduszy-cka-Machnikowa and Lech 1976; Lech 1979, 1983; Szeliga 2007; Szeliga et al. 2021). This may be evidenced by the occurrence of the refitted fragments of blades from Pit 40 (Figure 11: 7). On the other hand, due to the high fragility of obsidian, it cannot be ruled out that the high frequency of fragments in the assemblage could (at least in part) reflect the damage caused during the use of what were originally entire specimens.

The surviving whole specimens tend to be very small (Figure 4). Their length ranges from 15 to 37 mm, with the vast majority of specimens grouped in the range of 20–30 mm. This corresponds to the sizes of the cores (Table 2; Figure 3), as well as the blade negatives preserved within their flaking surfaces. The width of the blades varies between 3.5 and 13 mm, with specimens narrower than 5 mm and wider than 10 mm being extremely

Table 3. Opatów, Site 2. State of preservation of blades: C – complete specimens, PP – proximal parts, MP – median parts, DP – distal parts.

Blades	N	%
C	49	31.21
PP	54	34.39
MP	24	15.29
DP	30	19.11
Total	157	100.0

rare. In terms of thickness, the obsidian blades from Opatów are generally in the range of 0.5–4 mm (the most numerous are in the range of 1.5–2.5 mm).

The analysis of the nature of the upper face of the whole blades and their fragments revealed that the most numerous group are specimens with a dorsal surface that is completely scarred (e.g., Figures 8: 1, 11–14; 9: 1–2, 4, 16, 19–23; 10: 1, 3, 5–7, 10, 16). They come from the advanced phase of the exploitation of cores and are characterized by the orientation of the scar following the direction of knapping (Table 4). They are dominated by specimens with trapezoidal and triangular cross sections. The number of the remaining blades with the dorsal surface completely scarred—that is, specimens with flake scars perpendicular to the axis, or crested (Figure 7: 5–6)—is incidental (3 items, 1.91% of the total; see Table 4). The blade assemblage also contains a relatively small quantity of fully and partially cortical specimens (53 specimens in total; 32.71% of the total). There is only a slightly higher frequency of blades that are longitudinally cortical, which is the result of the expansion of the flaking surface onto the cortical sides of the cores. There are 27 examples of this, constituting 17.20% of all blades (e.g., Figures 7: 12, 16; 8: 4, 7, 15, 19; 9: 6–8, 12–13, 15, 18).

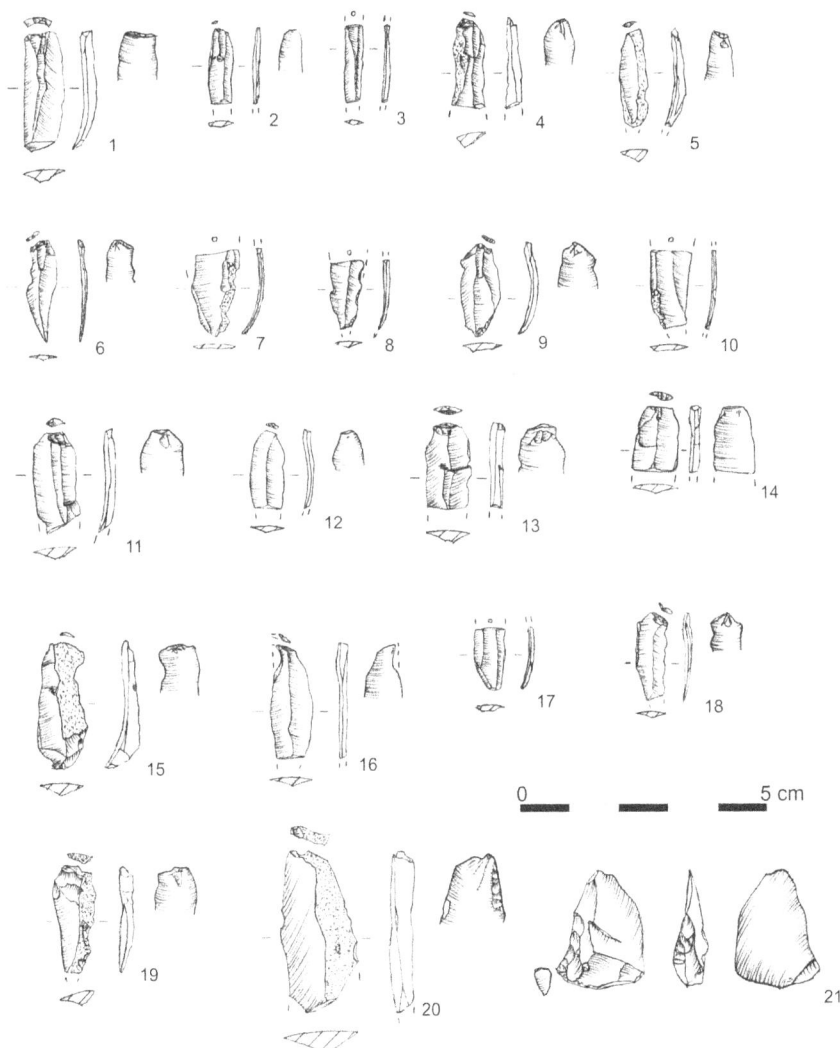

Figure 8: Opatów, Site 2. Selection of artifacts from Feature 35.

Figure 9: Opatów, Site 2. Selection of artifacts from Feature 37.

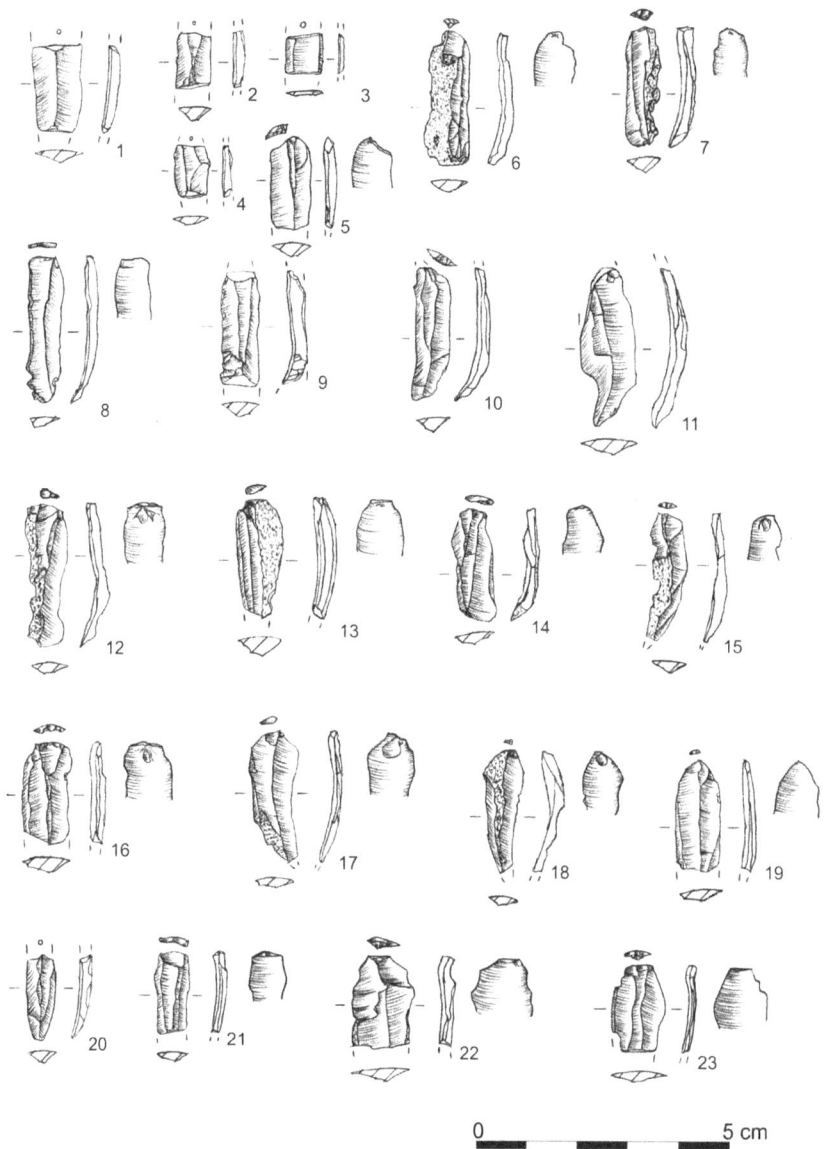

Table 4. Opatów, Site 2. Differentiation of the dorsal surfaces of blades, their fragments and tools made from blades.

Specimens	N	%
Cortical blades:		
a) completely	2	1.27
b) longitudinally	27	17.20
c) partially	24	15.29
Completely scar blades:		
a) negatives parallel to the axis	101	64.33
b) negatives perpendicular to the axis	3	1.91
Total	157	100.0

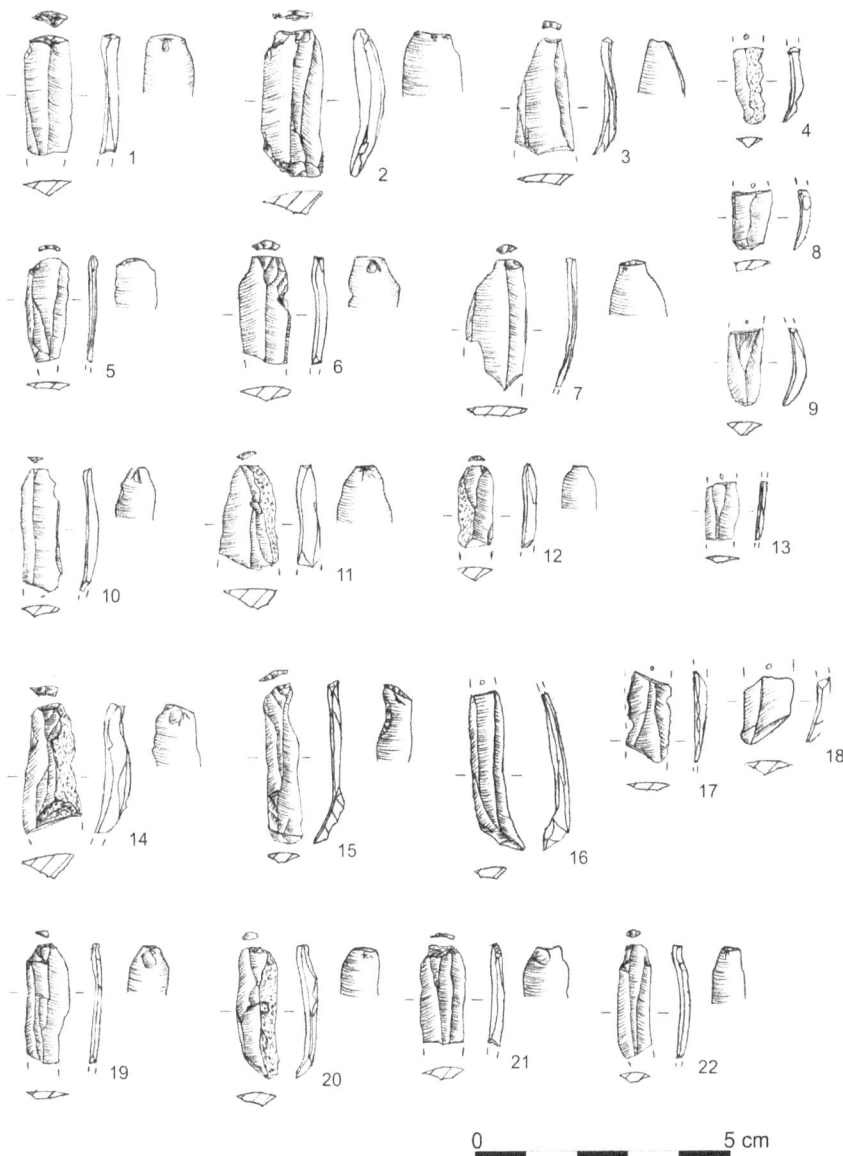

Figure 10: Opatów, site 2. Selection of artifacts from Feature 37.

The vast majority of the blades have one-, two-, or multi-negative butts—that is, butts formed by single or few removals and very often showing signs of faceting. This is reflected in the morphology of the butts and, more precisely, in their shape and height. The vast majority of the analyzed specimens have flat or slightly concave butts, with an outline that is most often broadly triangular or lenticular (e.g., Figures 8: 12, 14; 10: 1–2). On the other hand, there are very low frequencies of specimens with cortical and linear butts. These data from blade butts confirm the information provided by an examination of the cores about the widespread use of treatments aimed at the preparation of the striking platform and correction of the core angle during the use of the cores.

Table 5: Opatów, Site 2. Structure of items of the third morphological group.

Specimens	N	%
Flakes	48	58.56
Chips	34	41.44
Total	82	100.0

Table 6. Opatów, Site 2. Differentiation of the dorsal surfaces of flakes.

Specimens	N	%
Completely cortical flakes	6	12.5
Partially cortical flakes	24	50.0
Completely scar flakes	18	27.5
Total	48	100.0

Figure 11: Opatów, Site 2: Selection of artifacts from Features: 40 (1–7, 14), 35 (12), and 37 (13), as well as from trench 1924/1929 (8–11).

Flakes and chips

Flakes and chips are the second largest group of obsidian products in the analyzed inventory (82 specimens, 31.06% of the entire collection; Figure 2). More than half of the finds in this category are flakes (48 items; see Table 5), including mainly whole specimens (39 items). The remaining finds (34 items) are mostly microlithic chips—usually not exceeding 7 mm in length.

The analysis of the dorsal surfaces of the flakes shows a very high proportion of fully and partially cortical specimens (30 items, 62.5% of the total; see Table 6). They are the remains of both the initial preparation of the cores, as well as core rejuvenation actions (including mainly the striking platforms) during the exploitation (Figures 6: 4; 11: 12). The products of the process of the rejuvenation of the striking platforms of blade cores are also present among the specimens with completely scarred dorsal surfaces (Figure 6: 6; 11: 13). In a single case, there was a partially cortical flake (Figure 11: 14), which was struck from the striking platform of a blade core, removing part of its flaking surface. This specimen is most likely a remnant of an unsuccessful attempt to remove a blade following the orientation of the basic plane of exploitation of the core. However, it is also possible that it comes from an early stage of the intentional transformation of a blade core into a flake form.

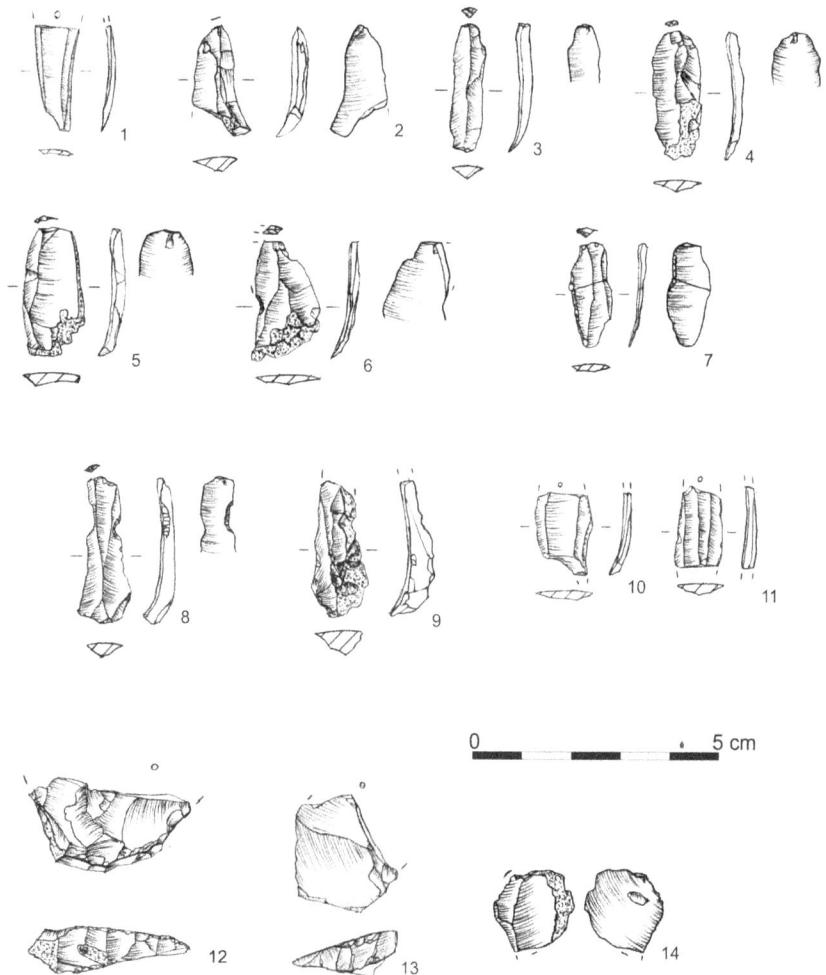

The analysis of the butts of the flakes indicates that the largest group consisted of specimens with the butts formed by a single or few strikes (23 items in total, 56.10% of the collection; see Table 7). This corresponds to the frequent use of core rejuvenation procedures, including core renewal, platform rejuvenation, as well as core reorientation, documented in the collection. Flakes with cortical butts (13 items, 31.71% of the total) constitute a slightly less numerous yet also an important group in the assemblage. These were recorded mainly in the case of platform rejuvenation flakes struck from the sides of the core. In turn, the specimens with linear butts constituted the lowest percentage in the set (Table 7).

Table 7. Opatów, Site 2. Differentiation of the butts of flakes.

Butts	N	%
Cortical	13	31.71
Plain	20	48.78
Facetted	3	7.32
Linear	5	12.20
Total	41	100.0

Retouched tools (implements)

The last category of obsidian finds in the assemblage are typological tools, represented by a total of 12 items, constituting 4.55% of the entire inventory (Figure 2). Among them, the most numerous group is retouched blades (7 items). In most cases, these are blades with backs that are completely scarred (with the orientation of the scars parallel with the direction of knapping), with very fine retouching. Retouching is located on the positive or negative side and includes straight (Figure 11: 7) or concave (Figures 10: 15; 11: 8) edges, a few or a dozen millimeters long. Only in two cases was retouching found on longitudinally cortical specimens (Figures 7: 16; 8: 20). The collection of prepared obsidian tools from Opatów is supplemented with five retouched flakes, represented by both fully cortical specimens, as well as platform rejuvenation flakes and ordinary, completely scarred flakes (Figure 8: 21). They show no regularities neither in terms of technological origins or sizes of individual specimens nor in terms of the location and method of retouching and the course of the retouched edge.

EDXRF analysis and results

In our analysis, 264 obsidian artifacts from Opatów were subjected to energy dispersive X-ray fluorescence (EDXRF) analysis using a QuanX EDXRF spectrometer, with instrumental setting and analysis conditions identical to those reported by Hughes and Werra (2014), Hughes and colleagues (2018), and Hughes and Ryzhov (2018).

Figure 12 presents Sr vs. Zr composition data for the 69 artifacts that were large enough to generate quantitative composition estimates, while Figure 13 is a ternary diagram plot of the relative proportions of Rb, Sr, and Zr in the other 195 Opatów artifacts that were too small for reliable quantitative analysis (i.e., those < 10 mm in diameter and < ca. 1.5 mm thick; see analysis protocol in Hughes 2010).

The quantitative composition data for large artifacts appear in Supplement A, and data for smaller specimens are listed in Supplement B. Figures 12 and 13 present the results in diagrammatic form, documenting that all artifacts analyzed were made from obsidian of the Carpathian 1 chemical type, which occurs in the Zemplén Mountains in southeast Slovakia. These results correspond very well with the results so far obtained

Figure 12. Sr vs. Zr composition of 69 large obsidian artifacts from Opatów. Dashed lines represent the range of variation of composition measured in archaeologically significant geological reference samples (adapted from Hughes and Werra 2014: Figure 5). The symbols plot the artifacts listed in Supplement A.

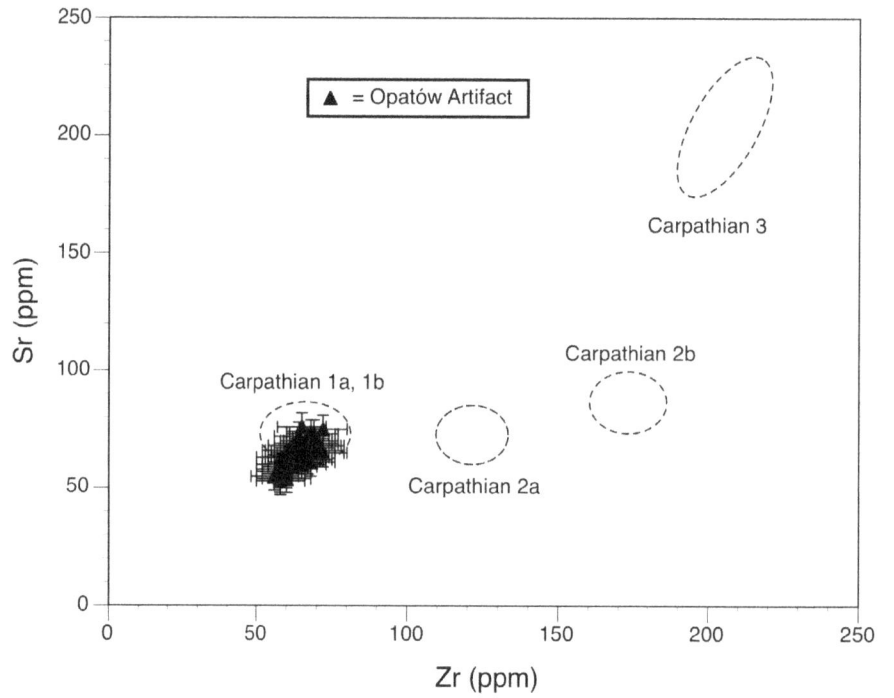

Figure 13. Normalized Rb/Sr/Zr composition of 195 small obsidian artifacts from Opatów. Dashed lines show the range of composition variation measured in archaeologically significant geological reference samples (adapted from Hughes and Werra 2014: Figure 5). The symbols plot the artifacts listed in Supplement B.

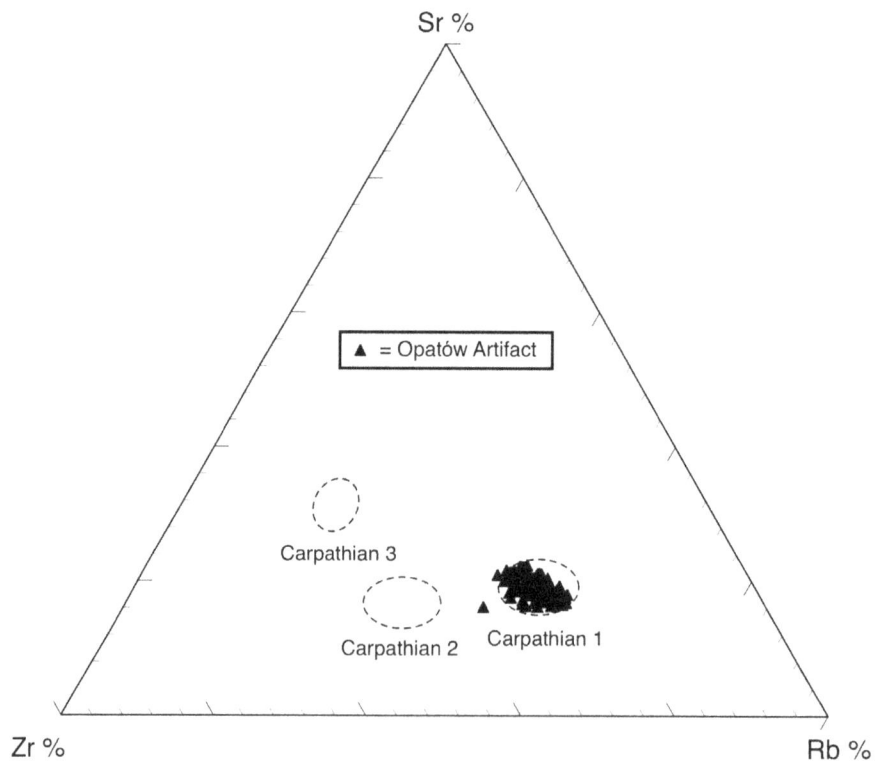

from adjacent areas (see Biró 2014; 2018; Constantinescu et al. 2014, 148; Kabaciński et al. 2015; Burgert et al. 2017; Riebe 2019; Szeliga et al. 2021; Werra et al. 2021). Therefore, no source-specific differences within or among different Opatów technological, morphological, or typological groups were identified by EDXRF analysis.

Summary

As stated above, EDXRF analysis did not reveal any source-specific differences within or among different morphological or typological groups. The morphometric characteristics of the obsidian products from site 2 in Opatów indicate that the processing of this material within the site was most likely concentrated only on the acquisition of very fine, even microlithic, blades as flake and blade blanks. This is indicated both by the presence of only blade core forms in the collection, as well as the dominant frequency of blades in the overall structure of the entire inventory (Figure 2). The flakes, which also constitute a large number of the finds, most often represent the residues of various preparatory and rejuvenation operations undertaken during the exploitation of blade cores (Figures 6: 4, 6; 11: 12–13).

The cores were exploited sparingly, as evidenced by the very rare use of preparatory treatments. This manifests itself both in the cores themselves, as well as in the low proportion of technical debris, i.e., the crested (Figures 7: 5–6; 11: 2), as well as the fully cortical blades and flakes (Tables 4 and 6). The core preparation mostly affected the striking platforms of the cores, and only sporadically their remaining parts. On the other hand, various types of rejuvenation procedures, undertaken during the exploitation of the cores, were used slightly more often. Maintaining an appropriate striking angle and leveling up of the platform edge enabled the effective production of a semi-raw material of appropriate dimensions and slenderness. The care for the most effective use of the cores is also confirmed by quite frequent changes in their orientation.

After knapping, blades and flakes were seldom processed into retouched tools, which allows us to conclude that they were the main product of the knapping process. Their very sharp edges made them suitable for various types of work—in particular, for cutting soft and medium-hard raw materials (Hurcombe 1992). The limited scope of use of obsidian tools may also be indicated by a very small degree of typological diversity of the prepared tool forms, represented only by retouched blades and flakes. Due to the microlithic size, as well as the high fragility of obsidian, these blades and flakes were probably used after prior hafting. An indirect confirmation of such a conclusion may be the clear preponderance in the assemblage of blade fragments over whole specimens (Table 3). This may indicate that the blades were deliberately broken before hafting in order to remove their most curved (top) and thickest (butt) parts. These hypotheses, however, require verification in the course of further use-wear analyses and experimental work.

The scope of obsidian treatment documented at Opatów, as well as the morphological structure of the collection and the frequency of individual product categories, does not differ to any significant extent from other Neolithic obsidian inventories known from sites on the northern side of the Carpathians (e.g., Milisauskas 1986; Michalak-Ścibor

1994; Szeliga 2007, 2009; Wilczyński 2010, 2014a, 2014b; Szeliga et al. 2019, 2021). Some differences are observable only on the metric level, especially noteworthy in the small dimensions of the cores and the semi-raw blade material obtained from them. The question of whether they are only a natural and unintended reflection of the small size of nodules (and cores?) delivered to the settlement, or a conscious and planned production goal (determined, for example, by practical needs or cultural conditions), unfortunately, remains without an unequivocal answer.

Conclusions

Nearly 100 years have elapsed since the first excavations at site 2 in Opatów and half a century since the publication of the lithic assemblages from this site (Więckowska 1971). Despite the passage of time and the layers of dust on the boxes with artifacts, this site still contains important knowledge about the past. We have tapped some of that information through instrumental analysis of the obsidian, which shows broad similarities between and among Middle Neolithic communities.

The EDXRF analysis of the Opatów artifacts confirmed the previously documented role of Slovakian obsidian outcrops in the economy of Neolithic communities of Central and Eastern Europe (Biró 2014). The technological and morphological analyses presented here show that the use of this raw material by prehistoric communities at site 2 in Opatów does not differ, to any significant extent, from other Neolithic obsidian inventories known from sites on the northern side of the Carpathians. Nevertheless, it is worth noting that the obtained flake and blade blanks were the desirable final product, rather than other tools (e.g., end-scrapers, burins, perforators) that are very typical and common at other sites of this period. The only recorded treatments are rare retouching and intentional breaking to obtain the desired straight fragment.

The lithic inventory from Opatów contains 5,791 items, including 508 tools. The assemblage at Opatów is dominated by "chocolate" flint artifacts (88%); 5% are made from obsidian, 2% from Jurassic-Cracow flint, and 5% are burned specimens for which the raw material could not be determined (Więckowska 1971). This raw material is considered to be one of the highest-quality siliceous rocks in East-Central Europe, as is evidenced by its extensive use by prehistoric communities from the Paleolithic to Bronze Age, and even up to the Early Iron Age (Werra and Kerneder-Gubała 2021). This abundance of very high-quality raw materials (the deposits of "chocolate" flint are located ca. 55 km from Opatów) immediately raises the questions of why, and for what purposes, obsidian was necessary. One can imagine that all the specimens were brought together with the first wave of "colonizers" from the south. The technological analysis showed that the processing of obsidian was very purposeful; the Neolithic communities at Opatów knew how to handle this raw material, achieve the intended forms, and use them as tools. Along with the strong influences of the Stroked Pottery Culture, it seems likely that for the Lengyel culture and the communities from the Upper Tisza river basin, the use of obsidian was one of the material elements that marked social distinctions within and among the Samborzec-Opatów group. These socially driven factors probably influenced the technology of processing this raw material, as well as the use of the flake and blade blanks as a finished tool. Such distinctions were not registered in other tools, which were made predominantly of "chocolate" flint.

Apart from purely technological questions, instrumental analysis of obsidian allows us to consider whether obsidian was necessary for those communities, if it was required to complete a tradition or ritual, or if it had a culture-forming value. Was it an element through which a connection could be made with the ancestors? Was it symbolic of one's homeland, carried forward to make the new place like home (Mateiciucová 2010; Burgert 2016)? Perhaps the obsidian tools were used for other purposes than tools made from the local raw material (see Małecka-Kukawka 2001; Pyzel and Wąs 2018; Szeliga et al. 2021). These differences could have been conditioned both by practical (functional) and non-utilitarian (symbolic, ritual) considerations.

We hope that the experimental and use-wear analysis we have conducted on the obsidian finds from Opatów will assist in discovering other material cultural linkages among prehistoric communities in the area. The old, dusty boxes of artifacts from Opatów allow us to ask new questions—from new perspectives—about status, social ranking, and the value of obsidian among the Neolithic communities in the first half of the fifth millennium BC.

Acknowledgments

The archaeological research described in this chapter was conducted within the research project No 2018/29/B/HS3/01540 entitled "Investigation of the Sources and Uses of Obsidian during the Neolithic in Poland," led by Dagmara H. Werra, funded by the National Science Centre, Poland.

Supplements:

- Supplement A. Quantitative composition determinations for large obsidian artifacts from Opatów, Site 2, Opatów district.
  Available online: https://escholarship.org/uc/item/75c689n2
- Supplement B. Normalized Rb/Sr/Zr composition of small obsidian artifacts from Opatów, Site 2, Opatów district.
  Available online: https://escholarship.org/uc/item/75c689n2

References

Biró, Katalin T.

2014   Carpathian Obsidians: State of Art. In *Lithic Raw Material Exploitation and Circulation in Prehistory. A Comparative Perspective in Diverse Palaeoenvironments*, edited by M. Yamada and A. Ono, pp. 47–69. ERAUL 138, Liège.

2018   More on the State of Art of Hungarian Obsidians. *Archeometriai Műhely* 15: 213–223.

Burgert, Pavel

2016   The Status and the Role of 'Chocolate' Silicite in the Bohemian Neolithic. *Archaeologia Polona* 56: 49–64.

Burgert, Pavel, Antoní Přichystal, Lubomír Prokes, Jan Petřík, and Simona Hušková

2017   The Origin and Distribution of Obsidian in Prehistoric Bohemia. *Bulgarian e-Journal of Archaeology* 7: 1–15.

Constantinescu, B., D. Cristea-Stan, I. Kovács, and Z. Szőkefalvi-Nagy

2014   Provenance Studies of Central European Neolithic Obsidians Using External Beam Milli-PIXE Spectroscopy. *Nuclear Instruments and Methods in Physics Research Section B: Beam Interactions with Materials and Atoms* 318: 145–148.

de Grooth, Marjorie. E. Th. de.

1981   Fitted Together Bandkeramik Flint. In *Staringia 6. Third International Symposium on Flint, 24–27.05. 1979, Maastricht*, pp. 117–118. Maastricht, Netherlands.

1990   Technological and Socio-Economic Aspects of Bandkeramik Flint Working. In *The Big Puzzle. International Symposium on Refitting Stone Artefacts*, edited by Erwin Cziesla, Sabine Eickhoff, Nico Arts, and Doris Winter, pp. 197–210. Verlag, Bonn, Germany.

Dzieduszycka-Machnikowa, Anna, and Jacek Lech

1976   *Neolityczne zespoły pracowniane z kopalni krzemienia w Sąspowie.* Wrocław-Warszawa-Kraków-Gdańsk.

Fiedorczuk, Jan

1992   Późnopaleolityczne zespoły krzemienne ze stanowiska Rydno IV 57 w świetle metody składanek. *Przegląd Archeologiczny* 43: 47–59.

2006   *Final Paleolithic Camp Organization as Seen from the Perspective of Lithic Artefacts Refitting.* Institute of Archaeology and Ethnology Polish Academy of Sciences, Warszawa.

Govindaraju, K.

1994   Compilation of Working Values and Sample Descriptions for 383 Geostandards. *Geostandards Newsletter, Special Issue* 18: 1–158.

Hofman, J. L.

1992   Putting the Pieces Together: An Introduction to Refitting. In *Piecing Together the Past: Applications of Refitting Studies in Archaeology*, edited by J. L. Hofman and J. G. Enloe, pp. 1–20. British Archaeological Reports, International Series 578, Oxford.

Hughes, Richard E.

2010   Determining the Geologic Provenance of Tiny Obsidian Flakes in Archaeology Using Nondestructive EDXRF. *American Laboratory* 42(7): 27–31.

Hughes, Richard E., and Sergey Ryzhov

2018   Trace Element Characterization of Obsidian from the Transcarpathian Ukraine. *Journal of Archaeological Science: Reports* 19: 618–624.

Hughes, Richard E., and Dagmara H. Werra

2014   The Source of Late Mesolithic Obsidian Recovered from Rydno XIII/1959, Central Poland. *Archeologia Polski* 59(1–2): 31–46.

Hughes, Richard E., Dagmara H. Werra, and Zofia Sulgostowska

2018   On the Sources and Uses of Obsidian During the Paleolithic and Mesolithic in Poland. *Quaternary International* 468: 84–100.

Hurcombe, L. M.

1992   *Use Wear Analysis and Obsidian: Theory, Experiments and Results*. Sheffield Archaeological Monographs 4, Department of Archaeology and Prehistory, University of Sheffield, J. R. Collis Publications, Sheffield, England.

Jędrzejczyk, Konrad J.

2013   *Tożsamość narodowa społeczeństwa polskiego po okresie zaborów. rozważania na przykładzie archeologii w Polskim Towarzystwie Krajoznawczym w latach 1906–1950*. Państwowa Wyższa Szkoła Zawodowa we Włocła, Włocławek.

Kabaciński Jacek, Iwona Sobkowiak-Tabaka, Zsolt Kasztovszky, Sławomir Pietrzak, Jerzy J. Langer, Katalin T. Biró, and Boglárka Maróti

2015   Transcarpathian Influences in the Early Neolithic of Poland: A Case Study of Kowalewko and Rudna Wielka Sites. *Acta Archaeologica Carpathica* 50: 5–32.

Kaczanowska, Małgorzata

1971   Krzemienne materiały kultur neolitycznych pochodzenia południowego z terenu Nowej Huty. In *Z badań nad krzemieniarstwem neolitycznym i eneolitycznym*, edited by Janusz K. Kozłowski, pp. 10–24. Polskie towarzystwo Archeologiczne, Muzeum Archeologiczne w Krakowie, Kraków.

1985   *Rohstoffe, Technik und Typologie der Neolithischen Feuersteinindustrien im Nordteil des Flussgebietes der Mitteldonau*. Państwowe Wydawnictwo Naukowe, Warszawa.

1990   Uwagi o wczesnej fazie kultury lendzielskiej w Małopolsce. *Acta Archaeologica Carpathica* 29: 71–97.

Kaczanowska, Małgorzata, and Janusz K. Kozłowski
1994   Zur Problematik der Samborzec-Opatów-Gruppe. In *Internationales Symposium* Über *die Lengyel-Kultur 1888–1988. Znojmo-Kravsko-Těšetice 3–7. 10. 1988*, edited by Pavel Koštuřík, pp. 85–103. Masarykova Univerzita v Brně pro Filozofickou Fakultu, Brno.

Kaczanowska, Małgorzata, Janusz K. Kozłowski, and Anna Zakościelna
1987   Chipped Stone Industries of the Linear Band Pottery Culture Settlements in the Nowa Huta Region. *Przegląd Archeologiczny* 34: 93–132.

Kadrow, Sławomir, and Anna Zakościelna
2000   An Outline of the Evolution of Danubian Cultures in Małopolska and Western Ukraine. *Baltic-Pontic Studies* 9: 187–255.

Kamieńska, Jadwiga
1967   Z badań nad kulturą lendzielską w Małopolsce. *Archeologia Polski* 12: 257–279.

Kamieńska, Jadwiga, and Janusz K. Kozłowski
1970   The Lengyel and Tisza Cultures. In *The Neolithic in Poland*, edited by Tadeusz Wiślański, pp. 98–105. Cambridge University Press, Wrocław.

Kosterski-Spalski, Władysław
1963   Historia Muzeum Świętokrzyskiego w Kielcach 1908–1939. *Rocznik Muzeum* Świętokrzyskiego 1: 21–44.

Kowalewska-Marszałek, Hanna
2004   Wczesna faza kultury lendzielskiej na Wyżynie Sandomierskiej. *Materiały Archeologiczne Nowej Huty* 24: 35–48.
2007   The 'Lengyel-Polgár' Settlement in the Sandomierz Upland: Microregional Studies. In *The Lengyel, Polgár and Related Cultures in the Middle/Late Neolithic in Central Europe*, edited by Janusz K. Kozłowski and Pál Raczky, pp. 431–448. Polish Academy of Arts and Sciences Kraków, Kraków.
2012   Linear Pottery and Lengyel Settlement Structures on the Sandomierz Upland (Little Poland): Continuity or Change? In *Siedlungsstruktur Und Kulturwandel in Der Bandkeramik: Beiträge Der Internationalen Tagung "Neue Fragen Zur Bandkeramik Oder Alles Beim Alten ?!" Leipzig, 23. Bis 24. September 2010*, edited by Regina Smolnik, pp. 284–294. Landesamt für Archäologie Sachsen, Dresden, Germany.

Kozłowski, Janusz K.

1966   Próba Klasyfikacji materiałów zaliczanych do kultury lendzielskiej I nadcisańskiej w Polsce południowej. *Archeologia Polski* 11: 7–27.

1989   The Lithic Industry of the Eastern Linear Pottery Culture in Slovakia. *Slovenská Archeológia* 37(2): 377–410.

2004   Problemy kontynuacji rozwoju pomiędzy wczesnym i środkowym neolitem oraz genezy „cyklu lendzielsko-polgarskiego" w basenie górnej Wisły. *Materiały Archeologiczne Nowej Huty* 24: 11–18.

Kozłowski, Janusz K., and Stefan K. Kozłowski

1977   *Epoka kamienia na ziemiach polskich.* Panstwowe Wydawnictwo Naukowe, Warszawa.

Kozłowski, Janusz K., and Anna Kulczycka

1961   Materiały kultury starszej ceramiki wstęgowej z Olszanicy, pow. Kraków. *Materiały Archeologiczne* 3: 29–50.

Kulczycka-Leciejewiczowa, Anna

2004   Kultura lendzielsko-polgarska w Polsce, jej identyfikacja i podziały taksonomiczne. *Materiały Archeologiczne Nowej Huty* 24: 19–23.

Kulczyński, Janusz, and Zygmunt Pyzik

1959   *Pradzieje ziem województwa kieleckiego. Przewodnik po wystawie archeologicznej.* Muzeum Świętokrzyskie, Kielce, Poland.

Lech, Jacek

1979   Krzemieniarstwo w kulturze społeczności ceramiki wstęgowej rytej w Polsce. Próba zarysu. In *Początki neolityzacji Polski południowo-zachodniej*, edited by Włodzimierz Wojciechowski, Bogusław Gediga, and Lech Leciejewicz, pp. 121–136. Polska Akademia Nauk, Wrocław.

1981   Materiały krzemienne z osad społeczności wstęgowych w Niemczy, woj. Wałbrzych. Badania z lat 1971–1972. *Silesia Antiqua* 23: 39–45.

1983   Flint Work of the Early Farmers. Production Trends in Central European Chipping Industries from 4500–1200 B.C. An Outline. *Acta Archaeologica Carpathica* 22: 5–63.

1997   Materiały krzemienne z osad społeczności wczesnorolniczych w Strachowie, woj. Wrocław. In *Strachów. Osiedla neolitycznych rolników na Śląsku*, edited by Anna Kulczycka-Leciejewiczowa, pp. 229–265. Werk, Wrocław.

2008   Materiały krzemienne społeczności kultury ceramiki wstęgowej rytej z Samborca, pow. Sandomierz. In *Samborzec. Studium przemian kultury ceramiki wstęgowej rytej*, edited by Anna Kulczycka-Leciejewiczowa, pp. 151–204. Institute of Archaeology and Ethnology Polish Academy of Sciences, Wrocław.

Małecka-Kukawka, Jolanta

1992 *Krzemieniarstwo społeczności wczesnorolniczych ziemi chełmińskiej (2 połowa VI – IV tysiąclecie p.n.e.).* Toruń, Poland.

2001 *Między formą a funkcją. Traseologia neolitycznych zabytków krzemiennych z ziemi chełmińskiej.* Wydawnictwo Uniwersytetu Mikołaja Kopernika, Toruń, Poland.

Mateiciucová, Ina

2008 *Talking Stones: The Chipped Stone Industry in Lower Austria and Moravia and the Beginnings of the Neolithic in Central Europe (LBK), 5700–4900 BC.* Dissertationes Archaeologicae Brunensis/Pragensesque 4, Brno, Czech Republic.

2010 The Beginnings of the Neolithic and Raw Material Distribution Networks in Eastern Central Europe: Symbolic Dimensions of the Distribution Of Szentgál Radiolarite. In *Die Neolithisierung Mitteleuropas. Internationale Tagung, Mainz 24. bis 26. Juni 2005/ The spread of the Neolithic to Central Europe. International Symposium, Mainz, June 24–26, 2005,* edited by D. Gronenborn and J. Petrasch, pp. 273–300. Römisch-Germanisches Zentralmuseum, Mainz.

Michalak-Ścibor, Jolanta

1994 Nowe źródła do znajomości klasycznej fazy kultury malickiej z Wyżyny Sandomierskiej (stanowisko 2 w Ćmielowie). *Sprawozdania Archeologiczne* 46: 31–81.

Milisauskas, Sarunas

1986 *Early Neolithic Settlement and Society at Olszanica.* Memoirs of the Museum of Anthropology, University of Michigan, Ann Arbor, Michigan.

Podkowińska, Zofia

1953 Pierwsza charakterystyka stanowiska eneolitycznego na polu Grodzisko I we wsi Złota, pow. Sandomierz. *Wiadomości Archeologiczne* 19: 1–53.

1968 Sprawozdanie z prac wykopaliskowych w Opatowie, woj. kieleckie, w 1965 roku. *Sprawozdania Archeologiczne* 19: 41–52.

Přichystal, Antonín

2013 *Lithic Raw Materials in Prehistoric Times of Eastern Central Europe.* Masaryk University Press, Brno, Czech Republic.

Pyzel, Joanna, and Marcin Wąs

2018 Jurrasic-Cracow Flint in the Linear Pottery Culture in Kuyavia, Chełmno Land and the Lower Vistula Region. In *Between History and Archaeology: Papers in Honour of Jacek Lech,* edited by Dagmara H. Werra and Marzena Woźny, pp. 181–194. Archaeopress, Oxford.

Rácz, Béla

2018   The Carpathian 3 Obsidian. *Archeometriai Műhely* 15: 181–186.

Riebe, Danielle J.

2019   Sourcing Obsidian from Late Neolithic Sites on the Great Hungarian Plain: Preliminary p-XRF Compositional Results and the Socio-Cultural Implications. *Interdisciplinaria Archaeologica* 10: 113–120.

Řídký, Jaroslav, Petr Květina, Harald Stäuble, and Ivan Pavlů

2015   What is Changing and When – Post Linear Pottery Culture Life in Central Europe. *Anthropologie* 53(3): 333–340.

Rosania, Corinne N., Matthew T. Boulanger, Katalin T. Biró, Sergey Ryzhov, Gerhard Trnka, and Michael D. Glascock

2008   Revisiting Carpathian Obsidian. *Antiquity Project Gallery* 82(318), http://www.antiquity.ac.uk/projgall/rosania318/. Access date: 16.01.2024.

Szeliga, Marcin

2007   Der Zufluss und die Bedeutung des Karpatenobsidians in der Rohstoffwirtschaft der Postlinearen Donaugemeinschaften auf den polnischen Gebieten. In *Lengyel, Polgar and Related Cultures in the Middle/Late Neolithic*, edited by Janusz K. Kozłowski and Pál Raczky, pp. 295–307. Polish Academy of Arts and Sciences, Kraków.

2009   Znaczenie obsydianu karpackiego w gospodarce surowcowej najstarszych społeczności rolniczych na ziemiach polskich. In *Surowce naturalne w karpatach oraz ich wykorzystanie w pradziejach i wczesnym* średniowieczu, edited by Jan Gancarski, pp. 287–324. Muzeum Podkarpackie w Krośnie, Krosno.

2021   The Inflow of Obsidian North of the Carpathians during the Neolithic – Chrono-Cultural Variability of Distribution. In *Beyond The Glass Mountains. Papers Presented For The 2019 International Obsidian Conference 27–29 May 2019, Sárospatak*, edited by Katalin T. Biró, and András Markó, pp. 69–94. Magyar Nemzeti Múzeum, Budapest.

Szeliga, Marcin, Zsolt Kasztovszky, Grzegorz Osipowicz, and Veronika Szilágyi

2021   Obsidian in the Early Neolithic of the Upper Vistula Basin: Origin, Processing, Distribution and Use – A Case Study from Tominy (Southern Poland). *Praehistorische Zeitschrift* 96(1): 19–43.

Szeliga, Marcin, Michał Przeździecki, and Artur Grabarek

2019   Podlesie, Site 6 – The First Obsidian Inventory of the Linear Pottery Culture Communities from the Połaniec Basin. *Archaeologia Polona* 57: 197–211.

Tomaszewski, Andrzej Jacek

    1986   Metoda składanek wytworów kamiennych i jej walory poznawcze. *Archeologia Polski* 31: 239–277.

Thorpe Olwen, W., S. E. Warren, and J. G. Nandris

    1984   The Distribution and Provenance of Archaeological Obsidian in Central and Eastern Europe. *Journal of Archaeological Science* 11: 183–212.

Wąs, Marcin

    2005   *Technologia krzemieniarstwa kultury janisławickiej*, Vol. 3. Monografie Instytutu Archeologicznego Uniwersytetu Łódzkiego, Łódź, Poland.

Werra, Dagmara. H., Richard E. Hughes, Marek Nowak, Marián Vizdal, and Lýdia Gačková

    2021   Obsidian Source Use within the Alföld Linear Pottery Culture in Slovakia. *Sprawozdania Archeologiczne* 73(1): 331–369.

Werra, Dagmara H., and Katarzyna Kerneder-Gubała

    2021   'Chocolate' Flint Mining from Final Palaeolithic up to Early Iron Age – A Review. In *From Mine to User: Production and Procurement Systems of Siliceous Rocks in the European Neolithic and Bronze Age. Proceedings of the XVIII UISPP World Congress (June 4–9, 2018, Paris, France). Vol. 10. Session XXXIII–1&2*, edited by F. Bostyn, F. Giligny and P. Topping, pp. 42–56. Archaeopress Archaeology, Oxford.

Więckowska, Hanna

    1971   Materiały krzemienne i kamienne z osad kultury ceramiki wstęgowej i trzcinieckiej w Opatowie. In *Z polskich badań nad epoką kamienia*, edited by Waldemar Chmielewski, pp. 103–183. Ossolineum, Wrocław-Warszawa-Kraków-Gdańsk.

Wilczyński, Jarosław

    2010   The Techniques of Obsidian Treatment on the Malice Culture Settlement of Targowisko 11, Lesser Poland. *Przegląd Archeologiczny* 58: 23–37.

    2014a   Krzemienny i obsydianowy inwentarz kultury ceramiki wstęgowej rytej ze stanowiska Brzezie 17, gm. Kłaj. In *Brzezie 17. Osada Kultury Ceramiki Wstęgowej Rytej. Via Archaeologica. Źródła z badań wykopaliskowych na trasie Autostrady A4 w Małopolsce Vol. 9*, edited by Agnieszka Czekaj-Zastawny, pp. 499–546. Krakowski Zespół do Badań Autostrad, Instytut Archeologii Uniwersytetu Jagiellońskiego, Instytut Archeologii i Etnologii PAN O/Kraków, Muzeum Archeologiczne w Krakowie, Kraków.

    2014b   Neolityczne materiały kamienne z wielokulturowego stanowiska 10, 11 w Targowisku, pow. Wielicki. In *Targowisko, Stan. 10, 11. Osadnictwo z epoki kamienia.Via Archaeologica. Źródła z badań wykopaliskowych na trasie Autostrady A4 w Małopolsce, Vol. 8*, edited by Albert Zastawny, pp. 459–534. Krakowski

Zespół do Badań Autostrad, Instytut Archeologii Uniwersytetu Jagiellońskiego, Instytut Archeologii i Etnologii PAN O/Kraków, Muzeum Archeologiczne w Krakowie, Kraków.

2016   Flint, Obsidian, and Radiolarite in Lithic Inventories of the LBK Culture in Lesser Poland. In *Something Out of the Ordinary? Interpreting Diversity in the Early Neolithic Linearbandkeramik and Beyond*, edited by Luc Amkreutz, Fabian Haack, Daniela Hofmann, and Ivo Van Wijk, pp. 123–139. Cambridge Scholars Publishing, Newcastle upon Tyne.

Zakościelna, Anna

1996   *Krzemieniarstwo kultury wołyńsko-lubelskiej ceramiki malowanej*. Lubelskie Materiały Archeologiczne vol. 10, Lublin, Poland.

Zápotocká, Marie

2004   Chrudim. Příspěvek ke vztahu české skupiny kultury s vypíchanou keramikou k malopolské skupině Samborzec-Opatów. *Archeologické rozhledy* 56: 3–66.

CHAPTER 5

Imports and Outcrops: Characterizing the Baantu Obsidian Source and Artifacts from Mochena Borago Rockshelter, Wolaita, Ethiopia, Using Portable X-Ray Fluorescence

BENJAMIN DANIEL SMITH, LUCAS R. M. JOHNSON, STEVEN A. BRANDT

Benjamin Daniel Smith Coastal Carolina University, Conway, South Carolina, USA
Lucas R. M. Johnson Far Western Anthropological Research Group, Inc., 1180 Center Point Dr., Suite 100, Henderson, NV 89074, USA
Steven A. Brandt University of Florida, Gainseville, Florida, USA

Abstract

We characterized 42 obsidian samples from the Baantu obsidian source in southwestern Ethiopia, including 25 outcrop samples and 17 surface artifacts, using portable X-ray fluorescence (pXRF) spectroscopy. We then compared these source data to 116 obsidian artifacts from Mochena Borago Rockshelter. Results indicate that at least three geo-chemical source clusters are represented at the Baantu source: one derived from sampled outcrops and two deriving from as-yet unknown source locations. Comparing these data to obsidian artifacts at Mochena Borago excavated from levels dated to > 50 ka and ~44 ka BP, early levels dating to > 50 ka preserve obsidian from as many as six as-yet unidentified sources, while Baantu obsidians were in the minority. By ~44 ka cal BP, Mochena Borago occupants procured most, if not all, of their obsidian from the Baantu source. Comparison to regional published obsidian source data suggests little, if any, procurement from northern sources within the Ethiopian Rift. We need more regional survey and artifact characterizations to identify the spatial scale and directionality of stone procurement in this area, but these data provide evidence that the occupants of Mochena Borago Rockshelter engaged with a variety of stone raw materials across periods of major ecological and likely social change in the Late Pleistocene Horn of Africa.

Introduction

The Horn of Africa is one of the posited routes by which *Homo sapiens* populations dispersed across and eventually out of Africa during the Late Pleistocene (128–14 ka BP; see Beyin 2006; Groucutt et al. 2015; Shanahan et al. 2015; Stewart and Jones 2016). Obsidian sourcing can reveal behavioral patterns, particularly of material procurement and transport, during important periods of human evolutionary and demographic change. However, the relationship between lithic procurement at obsidian sources and consumption at archaeological sites remains poorly understood in most time periods due to the small number of geochemically characterized obsidian sources and artifactual obsidian assemblages in this part of the world.

The Baantu obsidian source lies ~18 km southeast of the city of Sodo in the Wolaita Zone (SNNPR) of southern Ethiopia (Figures 1 and 2). It is ~1–2 km^2 in surface area and composed of obsidian outcrops, exposed primarily by seasonal erosion, and extensive scatters of surface artifacts. Early and Middle Stone Age artifact types on the Baantu surface suggest that local lithic production was established by the late Middle Pleistocene (de la Torre et al. 2007; Schepers 2019). Today, local craft specialists use Baantu obsidian to produce hide scrapers, and others excavate this obsidian for household construction materials.

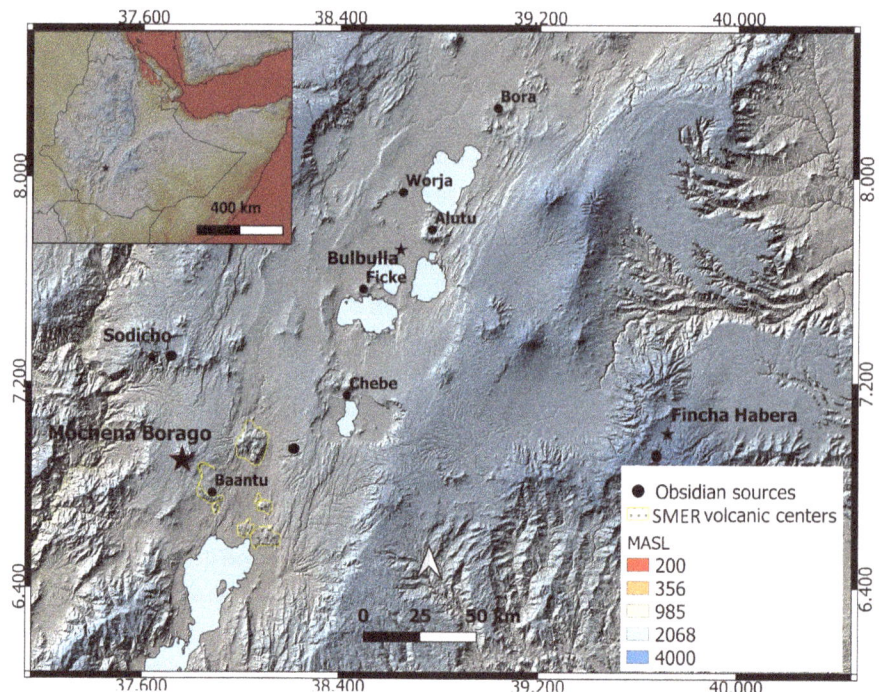

Figure 1. Study area including Gadamotta-area/Ziway-Shalla obsidian sources analyzed by Shackley and Sahle (2017). Inset shows location of Mochena Borago Rockshelter. Basemap from the MERIT DEM (Yamazaki et al. 2017).

Baantu was also a major source of toolstone for the occupants of Mochena Borago Rockshelter from > 50 ka to ~1,500 BP (located ~20 km northwest of Baantu; see Lesur et al. 2007; Brandt et al. 2012, 2017, 2023; Negash 2022), and for those at Sodicho Rockshelter from ~27 ka BP onward (located ~60 km northwest of Baantu; Hensel et al. 2021). To understand the context for these kinds of rockshelter occupations in Africa, which may date to periods of extreme global aridity such as Marine Isotope Stage (MIS) 4 (~73.5–60 ka BP) and MIS 2 (28–14 ka BP; see Hessler et al. 2010; Shenahan et al. 2010), Brandt and colleagues (Fisher 2010; Brandt et al. 2012) have posited that during arid periods of the Late Pleistocene, hunter-gatherers retreated into the southwestern Ethiopian Highlands, an area whose stable rainfall would have sustained ecological refugia. Although we do not aim to test this hypothesis per se, we discuss below the extent to which obsidian source data can be used to refine such a hypothesis at the local and regional scales.

Archaeologists have used energy dispersive X-ray fluorescence (EDXRF) spectroscopy to characterize trace elements in Ethiopian obsidians for many years (Kurashina 1978; Negash and Shackley 2006; Negash et al. 2006, 2007, 2010, 2011; Shackley and Sahle 2017; Zena et al. 2021). However, to our knowledge, only one published study used a portable X-ray fluorescence (pXRF) instrument (Arthur et al. 2019). Several authors have discussed issues surrounding the precision, validity, and uses of trace elemental data generated by pXRF (Frahm 2013a,b; Speakman and Shackley 2013). However, pXRF analyses have improved dramatically over the past 10 years as knowledge of regional obsidian geochemistry has grown and analysts have refined the analytical techniques involved (Frahm et al. 2017; Frahm 2019; Johnson et al. 2021).

In this study, we identify the range of variation in trace elements at the Baantu source from a variety of samples, including both outcrops and surface materials. We then use these data to identify variation in obsidian procurement at Mochena Borago Rockshelter. By analyzing both outcrops and surface materials at Baantu, we identify the geochemical variability within a major obsidian source. By comparing earlier and latter periods of obsidian use at Mochena Borago, we identify a shift from diversity to consistency in obsidian use during the Late Pleistocene.

Regional geology and research background

The Main Ethiopian Rift (MER), subdivided into Northern, Central, and Southern sectors (NMER, CMER, and SMER), is a symmetrical graben system in which Late Oligocene to Late Pleistocene (~32–0.13 ma) volcanic eruptions overlay older crystalline basement rocks. In some areas, lavas from these eruptions have extruded through Pleistocene alluvial sediments (WoldeGabriel et al. 1990: 442). Negash and colleagues (2020) geochemically characterized 45 Ethiopian obsidians using electron microprobe analysis (EPMA), as well as thin-section petrography. They found that most were rhyolitic, while a few were trachyte and dacite. In the SMER, east of Sodo, at least nine separate volcanic centers (Figure 1; see Chernet 2011) range in age from Late Pliocene to Late Pleistocene. Those closest to Mochena Borago and Baantu include Mt. Damota (where Mochena Borago is located, ~2.9 ma; WoldeGabriel et al. 1990), the Hobicha Caldera (~1.57 ma; Chernet 2011), and

Mt. Duguna Fango (~460-430 ka; Bigazzi et al. 1993). Hobicha and Duguna Fango are part of the Wonji Group fault system, which consists of a variety of mostly rhyolitic lava flows, pyroclastic rocks, and volcanoclastic sediments dating to < 1.6 ma (Kazmin 1979; WoldeGabriel et al. 1990). Baantu Hill was formed by lava flows and other volcanoclastic materials protruding from the southern inner rim of the Hobicha Caldera. What we call the Baantu obsidian source area is centered on a large erosional area on the northern face of Baantu Hill. It includes both outcrops and obsidian lithic artifacts located on, and in some cases completely covering, the surface of the hill slope (Figure 2).

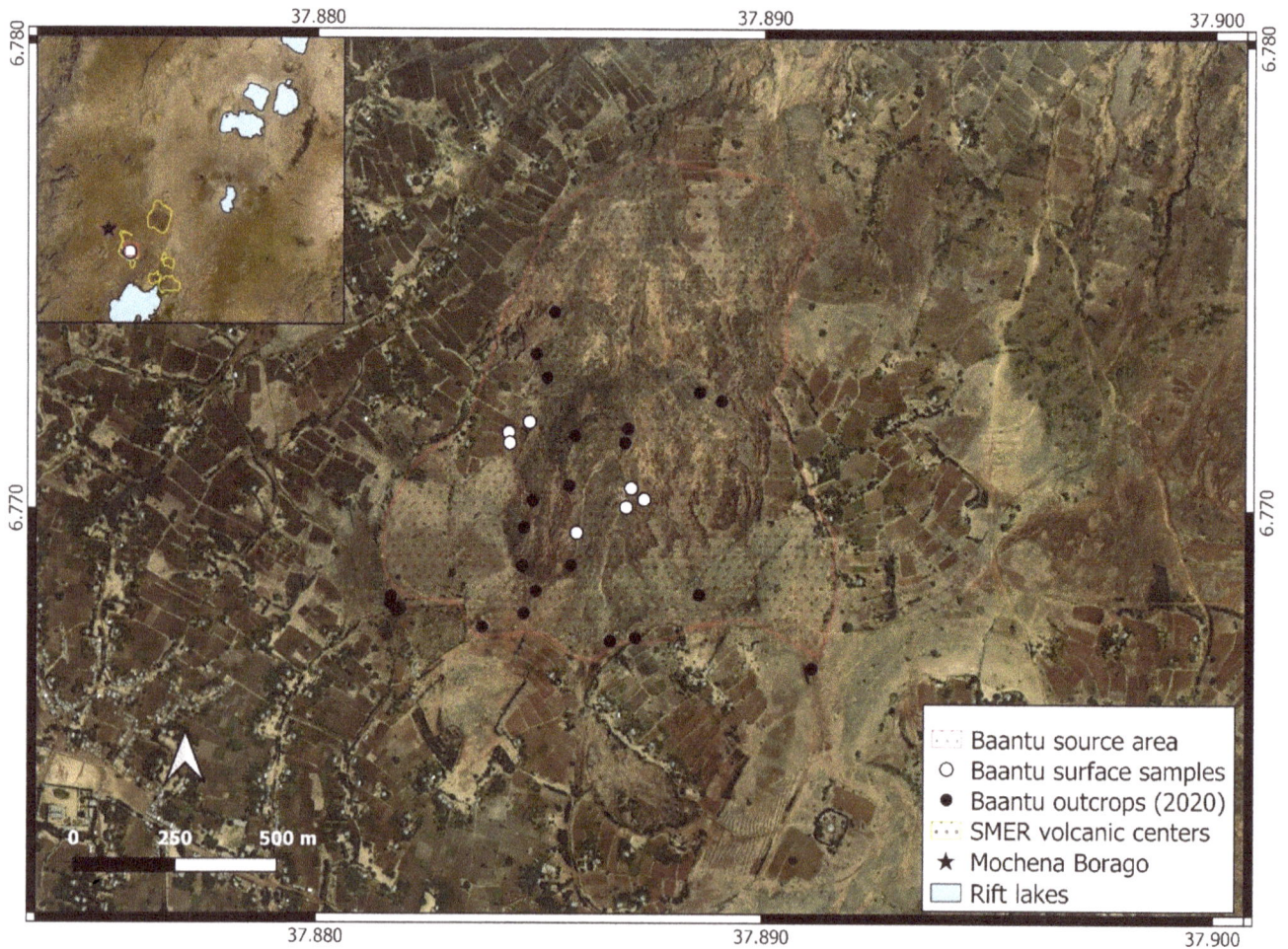

Figure 2. The Baantu source area. The top of Baantu Hill is in the SE and the Bisare River is NE. Black dots are obsidian outcrop samples, and white circles represent surface collections. Inset represents study area shown in Figure 1. Basemaps from ESRI Satellite (ArcGIS/World Imagery).

Seasonal tributaries of the Bisare River, which flow eastward into the Bilate River, have exposed obsidian outcrops at the Baantu source. The sediment profiles revealed by this erosion resemble those described elsewhere in the Hobicha Caldera by Benito-Calvo and colleagues (2007). These include a Lower Unit of yellowish coarse grain sediments, a Middle Unit of red-brown silts, sands, and clays (these clays are mined today at Baantu for pottery production; see Figure 3), and an Upper Unit of white sands and soils. At Baantu, the Middle Unit also contains erosional contacts, buried soils, pyroclastic sediments, and rich lenses of obsidian artifacts (mostly the results of primary-stage reduction but also many cores and some shaped bifaces; see Figure 4). We have not observed lithics in the Lower Unit at Baantu.

Figure 3. The Baantu source area including A) obsidian outcrops, B) surface lithics, and C) Middle Unit (*sensu* Benito-Calvo et al. 2007) red-brown clays mined today for pottery production.

Figure 4. Baantu stratigraphy. White bars on the right delineate Lower Unit yellowish coarse sediments, Middle Unit red-brown silts/clays, and Upper Unit white sands and plow zone soils (Benito-Calvo et al. 2007). Arrows point to obsidian artifacts buried at Lower/ Middle Unit contact and Middle/Upper Unit contact. Olivia Kracht (163 cm) for scale, goat's height unknown.

According to locals, "Baantu" was a previous Oromo owner of Baantu Hill. In the Oromo language, the long "aah" sound is typically spelled with two As, so here we prefer this spelling over "Bantu" (Negash et al. 2020; Negash 2022). Baantu contains obsidian in two contexts: A) outcrops exposed either by erosion or mining and B) surface materials including naturally eroded obsidians, quarrying debris, and artifacts. All toolstone-quality outcrops identified for this study have seen intensive quarrying and some removal of sediments to access toolstone. Some surface materials were also clearly recycled and reused, as many artifacts preserve evidence of more recent flake removals over neocortex (Figure 8).

De la Torre and colleagues (2007) identified 23 ESA-LSA sites in one of the few archaeological surveys of the area. They described one "Acheulean" site (A-9) eroding from the lower Middle Unit. The other sites contained artifacts typical of Middle or Later Stone Age contexts. Schepers (2019) has also discussed a "twisted" biface manufacturing strategy appearing throughout the Hobicha area. Large bifaces like these are typically the most heavily patinated and devitrified obsidian artifacts on the surface at Baantu. Other surface artifacts include a wide variety of Levallois cores. These include "classic" preferential types, recurrent Levallois, and "Nubian" Levallois cores, types 1 and 2 (*sensu* Van Peer 1992, Rose et al. 2011). East of the Baantu source area on Baantu Hill, there are also recently abandoned habitation sites, which are likely those of hideworkers who still visit the area today to collect raw materials.

In 2008, the Southwest Ethiopia Archaeological Project (SWEAP) began collecting Baantu obsidian samples for pXRF characterization. In 2014 SWEAP excavated a 2 x 2 meter unit immediately west of the northward eroding sediments, described above. This excavation unit cut through the Upper Unit plow zone and the upper levels of the Middle Unit. Excavators uncovered dense concentrations of obsidian artifacts, which were mostly the results of primary stage lithic reduction, as well as a few more intensively reduced cores and rare shaped tools. Based on nearby erosional profiles, this upper lens appears to be one of several stratified within the Middle and Upper Units (Figure 4). Deeper lenses appear several meters into the Middle Unit and likely represent some of the oldest episodes of Baantu obsidian exploitation in the area.

Obsidian sourcing studies and Late Pleistocene obsidian procurement in Ethiopia

Many sourcing techniques have been used to characterize major, minor, and trace elements in obsidian from the Horn of Africa (Cann and Renfrew 1964; Muir and Havernel 1974; Kurishina 1978; Clark et al. 1984; Francaviglia 1990; Bavay 2000; Negash et al. 2006, 2007, 2010, 2011, 2020; Negash and Shackley 2006; Glascock et al. 2008; Ménard et al. 2014, Shackley and Sahle 2017; Arthur et al. 2019; Oppenheimer et al. 2019; Khalidi et al. 2020; Negash 2022). Studies using XRF spectroscopy have clearly identified inter-source variation between eastern African obsidian sources by quantifying elements in the mid-Z range—particularly, Y, Nb, and Ba (Brown and Nash 2014; Shackley and Sahle 2017). However, as Shackley and Sahle (2017) note, there can be a high degree of overlap within these elements—particularly within areas of geographically overlapping and temporally confined volcanism, such as the SMER. Source standards employed in XRF calibrations, many of which are still derived from North American obsidian, need to be carefully evaluated or augmented when attempting to quantify obsidian geochemistry in these regions (see also Frahm 2019). In some cases, sources can be differentiated through methods combining both geochemical fingerprinting and forensic $^{40}Ar/^{39}Ar$ dating (Vogel et al. 2006) or magnetic susceptibility (McDougal et al. 1983; Hillis et al. 2010).

Long-distance stone transport and regionalization in manufacturing traditions are hallmarks of the so-called Middle Stone Age and the emergence of modern human behavioral complexity and variability in Africa during the Middle and Late Pleistocene (Clark 1988; McBrearty and Brooks 2000). Obsidian sourcing studies across eastern Africa have shown that Early-Middle Pleistocene obsidian procurement (~1.7–0.3 ma) was primarily localized and directed at certain preferred sources (Negash et al. 2006; Piperno 2009; Shackley and Sahle 2017). Middle-Late Pleistocene sites, especially those dating to < 100 ka BP, often preserve higher proportions of obsidian and a mix of localized and long-distance obsidian transport (Negash and Shackley 2006: Negash et al. 2007, 2010; Ambrose 2012; Hensel et al. 2021).

McBrearty and Brooks (2000) have suggested that increased transport distances exceeding 300 km likely indicate exchange rather than direct procurement of materials. However, a relatively high proportion of more distant materials (that is, relative to locally available sources) may also indicate prolonged rather than down-the-line exchange (Ambrose 2001). In Ethiopia at the Late Pleistocene site of Porc Epic, Negash and Shackley (2006) identified obsidian transport over 250 km, and transport distances exceeding 250 km have also been identified in Kenya (Blegan et al. 2017: Brooks et al. 2018). However, at many sites dating to the Late Pleistocene and Early Holocene, obsidian procurement seems to have been highly localized (Brandt 1982, 1986; Negash et al. 2007; Ménard et al. 2014; Ossendorf et al. 2019; Hensel et al. 2021). To understand the circumstances under which direct procurement or trade might have been preferred, we must identify available obsidian sources and explore the occupation histories recorded at deeply stratified sites.

Mochena Borago Rockshelter

Mochena Borago is a large rockshelter on the southwestern slopes of Mt. Damota, a trachytic volcano on the western margin of the SMER dated to ~2.9 ma (Figure 1; Wolde-Gabriel 1990). In the 1990s, a French team investigating early food production excavated the Holocene deposits at the site (Gutherz et al. 1998; Gutherz 2000; Lesur et al. 2007). Since 2006, SWEAP has excavated the Pleistocene deposits; and between 2009 and 2014, SWEAP was joined by members of the University of Cologne's Collaborative Research Center (Brandt et al. 2012, 2017, 2023).

The French team dated the Holocene deposits to between ~6 ka and 1,500 BP (Lesur et al. 2007). SWEAP later identified and dated several lithostratigraphic units (LSU) below these deposits, naming these "groups." SWEAP dated the LSU below the Holocene deposits to at least 36.6 ka BP (Brandt et al. 2017), suggesting a major unconformity in the depositional sequence. The Late Pleistocene stratigraphic sequence of LSU in this area of the shelter (called "MB1") was eventually dated to between > 49 and 39.6–36.6 ka (Bayesian model BP age ranges) via 56 radiocarbon dates (Figure 5; see Brandt et al. 2017). The 35 artifacts from MB1 discussed in this study come from an LSU called "S-Group" in excavation unit H9, which has been dated to between 46,136-44,790 and 45,879-43,885 (Bayesian age range BP; see Brandt et al. 2017: 364; see Table 4; Figure 5).

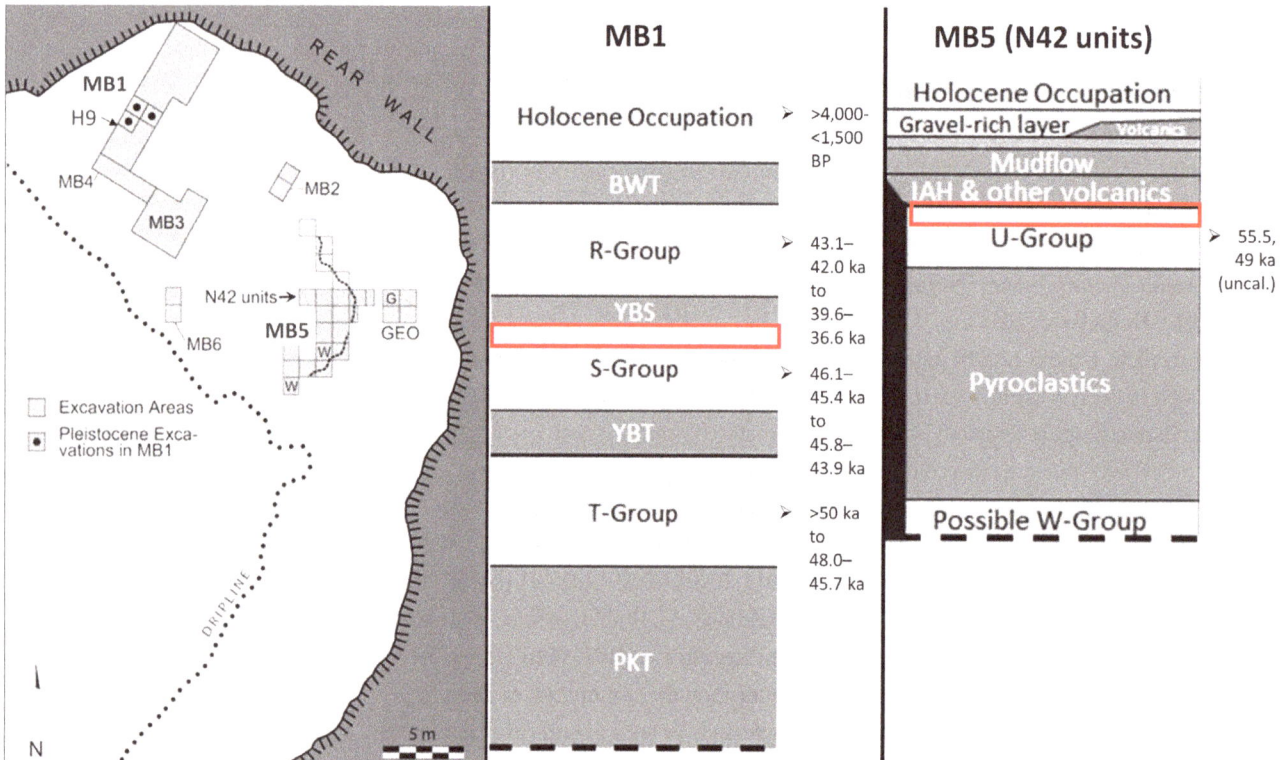

Figure 5. Mochena Borago excavation areas (left), and MB1/MB5 lithostratigraphic units (right) analyzed here (adapted from Brandt et al. 2023). Pleistocene MB5 dates are Bayesian Model BP age ranges (Brandt et al. 2017), while Holocene dates come from Lesur et al. (2007). Red rectangles indicate the location of analyzed samples.

Since 2012, SWEAP has focused excavations on southern areas of the shelter (Figure 5, "MB5"). These depositional sequences are more complex than those in MB1, with some variation between excavation units. The 87 MB5 artifacts analyzed in this study come from a single trench (N42) in the northern sector of MB5. A detailed review of the MB5 depositional sequence is published in Brandt et al. 2023. Radiometric dates for MB5 come from a dark brown stratum in N42 called ADR. It contained charcoal samples, which provided two uncalibrated C14 dates—the oldest dates at the shelter (> 55.5 ka BP and > 49 ka BP, Keck Carbon Cycle AMS Facility, University of California, Irvine).

Prior to this study, there was some evidence to suggest that occupants of Mochena Borago Rockshelter acquired their raw materials from different sources at different times, but no systematic attempt had yet been made to compare obsidian procurement between older and younger deposits at the site. Warren (2010) was the first to analyze obsidian artifacts from Mochena Borago, specifically from R-Group in MB1, using pXRF. Her results suggested that most of the MB1 obsidian artifacts were derived from Baantu obsidians. She also identified two unknown geochemical groups. Negash (2022) also identified two unknown geochemical groups while analyzing the Holocene materials from the French excavations using an electron probe micro-analyzer (EPMA). Finally, prior to the analyses in this study, excavators working in the deepest (and presumably oldest) levels of MB5 had noted relatively less obsidian overall.

Methods: sampling, EDXRF machine instrumentation, and data comparison

Trace element data collected in this study come from A) geological samples that Smith struck directly from Baantu obsidian outcrops in 2020, B) artifacts SWEAP collected from the Baantu surface between 2014 and 2020, and C) artifacts from Mochena Borago. We compared these data to other XRF data from sources north of the current area, specifically in and around the Gademotta caldera (Shackley and Sahle 2017).

SWEAP began analyzing Baantu surface samples using the project pXRF machine in 2008. Most of the Baantu surface samples analyzed here were collected in surveys organized by Ioana Dumitru in 2014. In 2020, Smith collected more surface materials along several transects centered around densely concentrated surface scatters and identified and sampled 32 obsidian outcrops (Figure 2).

The Baantu specimens we discuss in this chapter thus include 14 surface artifacts collected in 2014, 6 surface artifacts collected in 2020, and 25 outcrop samples collected in 2020 (Tables 2 and 4). Various SWEAP project members collected and analyzed the 2014 Baantu surface artifacts, while Smith analyzed 25 of the Baantu outcrop samples and 6 of the surface artifacts at the Ethiopian Authority for Research and Conservation of Cultural Heritage (ARCCH) in Addis Ababa in 2020.

All geological source materials and artifacts discussed in this study were analyzed on a single pXRF machine using an unmodified base-line obsidian calibration over a 12-year duration (2008–2020; for an overview, see Johnson et al. 2021). The instrument, a Bruker Tracer III-V +, serial number K0437, was equipped with a rhodium tube. The filter was composed of 1µm Ti, 12µm Al, 6µm Cu. It records MnKa1, FeKa1, ZnKa1, GaKa1, ThLa1, RbKa1, SrKa1, Y Ka1, ZrKa1, and NbKa1. The instrument was recalibrated on November 6, 2017, using the Bruker/MURR 40 standards, a LucasTooth regression method (Ferguson 2012; Johnson et al. 2021; Glascock 2020). Each geological and artifact sample was > 3mm thick, with a fresh break where possible, and large enough to cover the 3 x 4 mm X-ray spot size. Spectra were recorded once for each sample using Bruker's accompanying software S1PXRF at 40 kV in 180-second intervals. Several operators, mostly students, used the instrument over those 12 years; thus, some operator error may have influenced data recording (e.g., uncleaned surfaces, inadequate artifact placement) and subsequent interpretation of results (see below).

Various SWEAP project members, including Smith, analyzed MB5 artifacts from the levels 22–24 of units N42E35 and N42E35.0-5 between 2014 and 2018 (n = 87). These artifacts included both plotted finds and screened materials and consisted of a variety of debitage and shaped tools. Because various analysts collected the data, random sampling could not be guaranteed, but analyzed artifacts generally included a variety of shaped tools and debitage amenable to XRF characterization (see above). XRF spectra could not always be taken on a clean break, but all artifacts were washed, > 3mm in thickness, and large enough to cover the 3 x 4 mm X-ray beam.

Andrea Warren (2010) analyzed 35 artifacts from excavation unit H9 in MB1 (representing a 10% random sample of H9 artifacts from levels 37, 38, 39, and 45) in 2007–2008. However, when Warren analyzed these artifacts, the machine was in beta field testing, and spectra were recorded with different channel-element alignments. We reexamine these data in this study; and to compare these earlier artifact scans to the more recent assays of the Baantu source materials, Johnson and Smith extracted raw peak counts using the S1PXRF software version 3.8.30 after it was determined that raw counts could not be extracted using area of interest (AOI). This was done by placing a cursor within a given element peak at the precise keV location (e.g., Zr 15.7keV) and reading the gross peak photon count per second per each element from Mn to Nb. This count does not delete background noise (i.e., net count versus gross counts).

In sorting through the pre-2020 XRF data discussed here, Johnson identified a few errors. During the earliest period of analysis (2008–2010), the pXRF instrument used was being field tested, and, as a result, at least one source of error was introduced. The channel to energy alignment in the software was selected incorrectly or the software was not working correctly, which effects how raw spectral counts are converted to parts per million (ppm). This was corrected during the time Smith was collecting data and other studies were using this same instrument (Arthur et al. 2019). In addition to this source of error, the ppm conversion process was also being refined by Bruker Elemental. Because this instrument was one of the first to leave the factory, the original ppm calibration equation worksheet had to be refined slightly by updating the known values for the 40-standard set. It was corrected in 2017 by Lucas R. M. Johnson and Bruce Kaiser in the XRF lab at Far Western Anthropological Research Group and then returned to SWEAP for use in this study from 2018 onward. During the recalibration process, the USGS RGM-2 pressed pellet standard was analyzed 16 times to show the relative accuracy of the calibration on an independent international reference standard (Table 1). In addition, all prior spectra were converted to ppm through the revised calibration worksheet.

Results

Obsidian samples from Baantu, including outcrop samples and surface artifacts, reveal variability in the overall source signature that we attribute to the Baantu area (Figure 6; Table 2). Here, we identify at least three distinct geochemical groups at the Baantu source: the cluster revealed by the Baantu outcrops (38 samples, or 90%), Unknown Group 1 (3 samples, or 7%) and a single artifact representing Unknown Group 2 (2%). Previous analysts also misidentified a piece of black chert, which we include in Table 2 for comparison. The major differences between Baantu outcrop geochemistry and the unknown groups are in Rb, Y, Zr, and Nb content (Figure 6), reflecting patterns elsewhere in the rift (Brown and Nash 2014; Shackley and Sahle 2017).

Table 1. Comparison between USGS RGM-2 Reference Standard and Concentrations (ppm). Measured on Quant'X and Bruker Tracer III-SD EDXRF Instruments. Table by Johnson.

SAMPLE	Mn	Fe	Zn	Ga	Rb	Sr	Y	Zr	Nb	Ba	Ti	OBSIDIAN SOURCE
RGM-2 (recommended)[a]	273 ±8	13,008 ±273	33 ±2	16 ±1	147 ±5	108 ±5	24 ±2	222 ±17	9 ±0	842 ±35	0.15 ±0.01	Glass Mtn., CA
RGM-2 (n=1) (measured)[b]	246 ±16	12,546 ± 56	40 ±3	15 ±1	149 ±3	107 ±3	28 ±2	226 ±4	12 ±3	884 ±31	1,580 ±51	Glass Mtn., CA
RGM-2 (n=16) (measured)[c]	209 ±82	11,750 ± 312	33 ±12	18 ±2	139 ±3	95 ±6	25 ±3	200 ±6	9 ±2	nm	nm	Glass Mtn., CA

Notes: [a] USGS RGM-2 (Wilson 2009); [b] Measured on an EDXRF Quant'X, care of University California Berkeley Archaeological Research Facility in 2014; [c] Measured on Bruker Tracer III-V K-0437, average of 16 runs at 90 seconds each. nm = not measured.

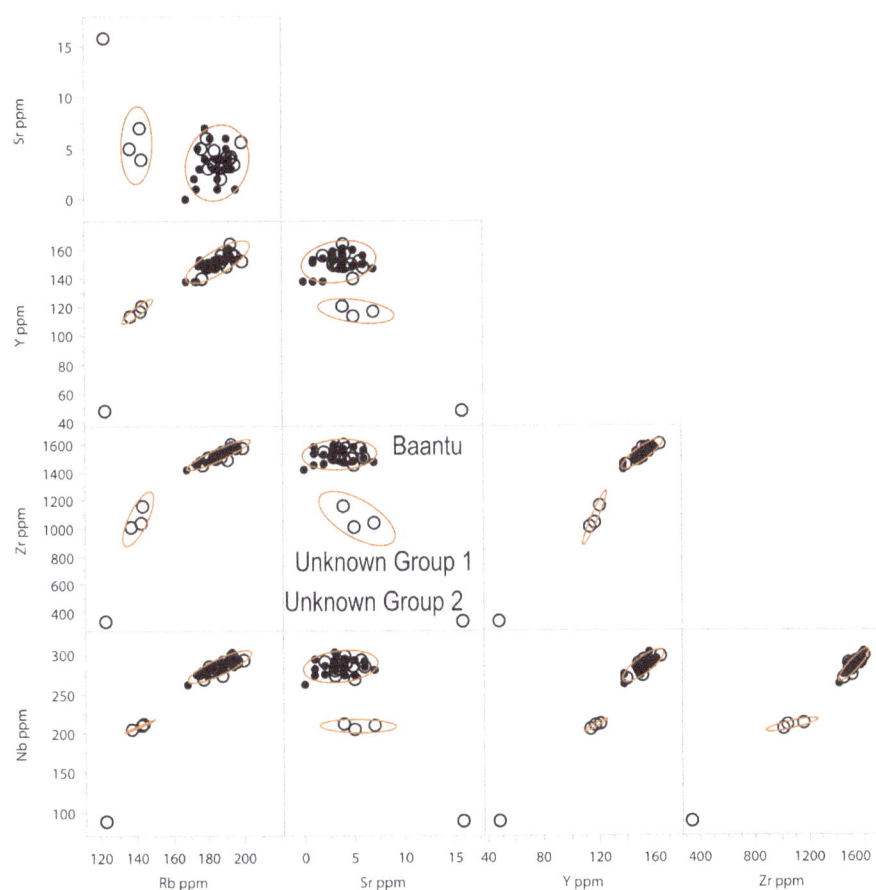

Figure 6. Element scatterplot matrix of all materials collected at the Baantu source. Outcrop samples are black dots and surface collections are open circles. 95% confidence ellipses were generated from the Baantu outcrops and surface materials, the latter of which preserve two as-yet unknown source groups. Figure by Smith and Johnson.

Table 2. Summary of ppm for all specimens collected at the Baantu source, including one piece of black chert.

Geo-chemical Group	Sample Location	Specimen	Mn	Fe	Zn	Ga	Th	Rb	Sr	Y	Zr	Nb
Baantu Outcrop	Outcrop	BA_4	1724	38470	268	29	37	187	4	154	1552	291
Baantu Outcrop	Outcrop	BA_5	1794	38786	273	30	42	185	4	147	1543	286
Baantu Outcrop	Outcrop	BA_6	1712	36001	252	29	38	179	4	150	1473	274
Baantu Outcrop	Outcrop	BA_7	1696	37832	253	28	41	188	4	152	1541	283
Baantu Outcrop	Outcrop	BA_8	1724	37310	270	29	34	184	3	153	1527	282
Baantu Outcrop	Outcrop	BA_9	1843	38009	253	31	39	190	3	152	1535	291
Baantu Outcrop	Outcrop	BA_10	1680	37036	240	31	42	186	3	153	1517	285
Baantu Outcrop	Outcrop	BA_11	1642	33598	238	30	38	168	0	138	1418	262
Baantu Outcrop	Outcrop	BA_12	1643	35679	260	29	38	173	2	138	1459	275
Baantu Outcrop	Outcrop	BA_13	1734	39444	270	30	40	193	3	156	1597	303
Baantu Outcrop	Outcrop	BA_14	1828	39475	292	32	40	192	3	159	1560	293
Baantu Outcrop	Outcrop	BA_15	1597	38410	253	31	41	186	2	154	1525	286
Baantu Outcrop	Outcrop	BA_17	1751	36660	249	29	37	175	5	149	1480	274
Baantu Outcrop	Outcrop	BA_18	1669	38953	263	30	42	192	3	156	1558	289
Baantu Outcrop	Outcrop	BA_20	1758	36754	255	30	40	176	3	153	1489	280
Baantu Outcrop	Outcrop	BA_21	1786	39154	262	28	39	196	1	153	1582	294
Baantu Outcrop	Outcrop	BA_22	1761	39128	265	32	38	191	4	157	1567	289
Baantu Outcrop	Outcrop	BA_23	1728	39289	274	30	40	192	4	161	1567	295
Baantu Outcrop	Outcrop	BA_24	1641	36119	255	30	37	186	1	151	1531	282

Table 2. Continued.

Baantu Outcrop	Outcrop	BA_25	1790	39362	262	28	37	191	5	160	1581	293
Baantu Outcrop	Outcrop	BA_31	1699	36838	248	29	39	179	7	147	1472	281
Baantu Outcrop	Outcrop	BA_32	1853	38770	265	31	41	191	6	156	1561	292
Baantu Outcrop	Outcrop	BA_33	1624	38184	264	33	42	182	6	150	1500	285
Baantu Outcrop	Outcrop	BA_35	1628	35246	250	28	37	174	1	138	1450	274
Baantu Outcrop	Outcrop	BA_37	1791	39181	291	31	36	194	3	154	1574	285
Baantu Outcrop	Surface	BA_26	1725	40760	273	30	42	188	2	156	1546	291
Baantu Outcrop	Surface	BA_27	1765	37554	262	30	43	180	6	148	1502	285
Baantu Outcrop	Surface	BA_28	1858	42778	298	30	38	193	4	164	1600	298
Baantu Outcrop	Surface	BA_30	2720	36478	226	28	38	177	5	140	1451	269
Baantu Outcrop	Surface	BN43-Area-A-Surface-Collection-2	1592	34969	267	24	26	188	3	151	1511	274
Baantu Outcrop	Surface	BN43-Area-A-Surface-Collection-3	1641	36120	261	26	30	184	4	149	1502	282
Baantu Outcrop	Surface	BN43-Area-A-Surface-Collection-4	1640	37055	289	24	25	199	6	152	1573	293
Baantu Outcrop	Surface	BN-23-Area-A-Surface-Collection-3	1631	34814	269	24	26	191	4	149	1488	289
Baantu Outcrop	Surface	BN-23-Area-A-Surface-Collection-4	1724	34953	261	20	31	184	3	149	1515	283
Baantu Outcrop	Surface	BN-23-Area-A-Surface-Collection-5	2528	37661	279	23	32	184	5	150	1497	286
Baantu Outcrop	Surface	BN-23-Area-A-Surface-Collection-6	1588	37282	291	26	29	196	3	156	1572	291

Table 2. Continued.

Baantu Outcrop	Surface	BN-27-Area-A-Surface-Collection-1	1671	38606	282	26	28	194	4	157	1573	290
Baantu Outcrop	Surface	BN-27-Area-A-Surface-Collection-2	1597	37748	296	28	28	181	3	149	1504	280
Black Chert	Surface	BN-23-Area-A-Surface-Collection-2	430	59928	12	15	3	7	4	9	67	11
Unknown Group 1	Surface	BA_19	1848	36894	206	31	33	137	5	114	1010	205
Unknown Group 1	Surface	BA_73	2150	39172	228	29	34	143	7	117	1040	210
Unknown Group 1	Surface	BN-23-Area-A-Surface-Collection-1	1769	34155	253	25	17	143	4	121	1160	212
Unknown Group 2	Surface	BN43-Area-A-Surface-Collection-1	1154	28272	99	22	11	122	16	48	337	88

We compared the Baantu outcrop data to Shackley and Sahle's (2017) data from the Gademotta area and used these data together to generate 95% confidence ellipses in JMP in two bivariate plots of Y/Rb and Y/Nb (Figure 7). We then added the MB5 artifact data to evaluate the percentage of Baantu vs. non-Baantu artifacts preserved in the levels sampled from U-Group in MB5.

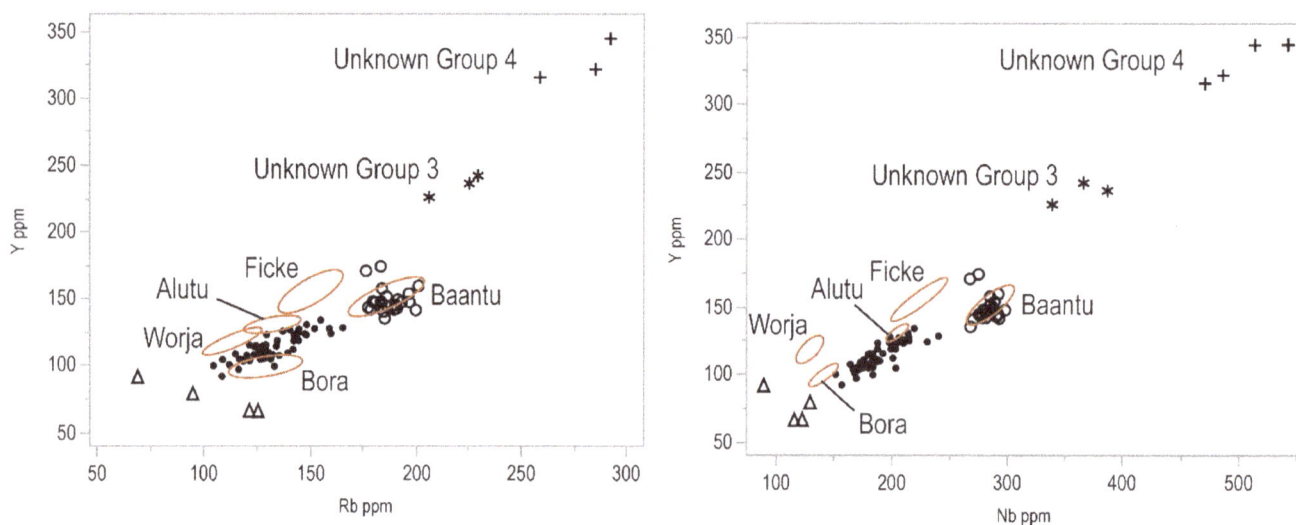

Figure 7. Two bivariate plots showing the distribution of MB5 artifacts. Red source ellipses were generated from the Baantu outcrops and from Alutu, Bora, Ficke, and Worja source data published in Shackley and Sahle (2017: Table 1). MB5 artifacts sourced to Baantu are open circles; Nonlocal artifacts are black dots; and Unknown Groups 3–7 are open triangles. Figure by Smith and Johnson.

In those older MB5 levels dated to beyond radiocarbon limits (> 50 ka), hunter-gatherers clearly procured some obsidian (22 artifacts, or 25%) from Baantu outcrops or surface materials derived therefrom. However, at least 65 artifacts (75%) come from up to 6 as-yet unidentified sources (see Figure 7; Table 3). These include the Nonlocal Group and Unknown Groups 3–7. None of these signatures match Unknown Groups 1 and 2 from the Baantu surface.

Table 3. Summary of ppm for MB5 specimens collected 2014–2018. Note that "Nonlocal" refers to geochemical signatures that likely correlate to sources north and east of Mochena Borago and Baantu, based on known relationships between Y and Rb and NB in the rift (Shackley, personal communication; see Figure 7).

Chemical Group	Specimen ID	Mn	Fe	Zn	Ga	Th	Rb	Sr	Y	Zr	Nb
Baantu	91602-N42E35.0-50-L23	1494	32566	257	24	27	185	4	136	1418	268
Baantu	91605-N42E35.0-50-L23	1675	35471	287	24	27	184	3	158	1551	285
Baantu	91615-N42E35.0-50-L23	1538	32923	259	23	29	186	4	145	1453	275
Baantu	91636-N42E35.0-50-L23 screened bag xrf#2	1574	33652	269	25	25	197	6	148	1500	281
Baantu	91649-N42E35-L24	1478	33035	264	21	25	184	5	148	1488	279
Baantu	91651-N42E35-L24	1579	34833	299	23	28	188	3	145	1521	292
Baantu	92556-N42E35.0 -L22-xrf #9	1356	32702	270	22	28	185	6	141	1457	270
Baantu	92556-N42E35.0 -L22-xrf#10	1445	34112	267	23	26	200	6	142	1467	293
Baantu	92556-N42E35.0 -L22-xrf#13	1881	42528	299	24	24	183	5	174	1442	275
Baantu	92556-N42E35.0 -L22-xrf#17	1435	32491	255	23	22	191	6	150	1483	288
Baantu	92572-N42E35.0-50-L23	1687	42599	313	23	24	177	7	171	1395	267
Baantu	92575-N42E35.0-50-L23	1588	34451	278	17	23	181	5	147	1485	285
Baantu	92579-N42E35.0-50-L23	1473	33751	283	24	28	190	4	142	1481	278
Baantu	92580-N42E35.0-50-L23	1519	36250	294	23	27	193	5	148	1528	298
Baantu	92583-N42E35.0-50-L23	1641	34582	278	25	28	186	3	152	1497	285
Baantu	92590-N42E35.0-50-L23	1284	36370	264	21	27	180	11	148	1489	288
Baantu	92807-N42E35.0-50-L23	1416	34264	305	23	25	178	13	144	1457	276
Baantu	92810-N42E35.0-50-L23	1486	36417	314	24	29	201	5	160	1552	292
Baantu	92835-N42E35.0-50-L23	1519	34356	269	22	22	191	5	143	1518	290

Table 3. Continued.

Baantu	92850-N42E35.0-50-L23	1532	34201	255	23	25	190	5	146	1476	277
Baantu	92861-N42E35.0-50-L23	1310	33948	262	25	28	179	8	142	1487	281
Baantu	92873-N42E35.0-50-L23	1708	35566	285	24	29	196	4	154	1548	289
Nonlocal	91604-N42E35.0-50-L23	1300	32857	210	24	14	116	9	97	974	170
Nonlocal	91613-N42E35.0-50-L23	1444	32368	239	22	17	128	3	105	1032	174
Nonlocal	91618-N42E35.0-50-L23	1525	32785	269	22	17	126	7	111	1034	176
Nonlocal	91619-N42E35.0-50-L23	1154	45028	274	24	23	109	8	105	1166	204
Nonlocal	91620-N42E35.0-50-L23	1624	31787	251	22	15	119	4	103	1003	172
Nonlocal	91622-N42E35.0-50-L23	1067	23032	183	24	19	130	4	107	981	165
Nonlocal	91636-N42E35.0-50-L23 screened bag xrf#1	1402	30690	211	24	15	112	6	101	941	169
Nonlocal	91636-N42E35.0-50-L23 screened bag xrf#4	1058	32888	240	26	20	142	5	124	1201	205
Nonlocal	91636-N42E35.0-50-L23 screened bag xrf#5	1349	31171	258	25	19	132	8	108	1051	182
Nonlocal	91641-N42E35-L24	1533	32117	234	25	16	128	3	115	1052	180
Nonlocal	91642-N42E35-L24	1793	35186	264	22	23	145	6	125	1225	215
Nonlocal	92556-N42E35.0 -L22-xrf#1	1566	32002	236	25	14	130	7	109	1068	186
Nonlocal	92556-N42E35.0 -L22-xrf#2	1185	33727	259	22	22	142	8	112	1160	201
Nonlocal	92556-N42E35.0 -L22-xrf#3	1522	31447	217	26	18	124	4	114	1053	184
Nonlocal	92556-N42E35.0 -L22-xrf#4	1569	34437	258	26	17	155	5	134	1236	219
Nonlocal	92556-N42E35.0 -L22-xrf#5pdz	1252	28147	199	23	16	109	3	92	915	158
Nonlocal	92556-N42E35.0 -L22-xrf#6	1408	31121	240	25	17	131	2	110	1059	190
Nonlocal	92556-N42E35.0 -L22-xrf#7	1112	34686	241	23	22	159	4	128	1131	199
Nonlocal	92556-N42E35.0 -L22-xrf#8	1004	31155	212	21	16	132	6	105	1076	184
Nonlocal	92556-N42E35.0 -L22-xrf#11	905	27946	177	23	14	104	12	100	883	152
Nonlocal	92556-N42E35.0 -L22-xrf#14	1442	29726	227	22	18	124	4	108	996	174
Nonlocal	92556-N42E35.0 -L22-xrf#15	1548	32709	221	20	19	140	6	110	1070	188

Table 3. Continued.

Nonlocal	92556-N42E35.0 -L22-xrf#16	1583	29244	198	24	16	120	4	108	990	176
Nonlocal	92556-N42E35.0 -L22-xrf#18	1605	34445	245	25	17	135	2	119	1089	188
Nonlocal	92556-N42E35.0 -L22-xrf#20	1433	32456	225	23	19	121	4	116	1036	183
Nonlocal	92565-N42E35.0-50-L23	1285	31087	253	25	21	140	4	127	1191	214
Nonlocal	92568-N42E35-L23	1667	33082	225	27	20	134	6	115	1102	181
Nonlocal	92574-N42E35-L23	1377	31410	266	22	22	152	5	128	1244	215
Nonlocal	92585-N42E35.0-50-L23	1298	30703	271	22	22	166	4	129	1285	240
Nonlocal	92586-N42E35.0-50-L23	1469	31616	213	23	14	115	5	109	998	174
Nonlocal	92592-N42E35.0-50-L23	1469	31695	224	22	16	131	5	109	1044	183
Nonlocal	92594-N42E35.0-50-L23	676	30789	177	22	18	130	10	123	1065	188
Nonlocal	92596N42E35.0-50-L23	1551	30282	226	21	14	122	4	104	952	167
Nonlocal	92599-N42E35.0-50-L23	1426	32328	241	25	20	136	5	108	1024	174
Nonlocal	92809-N42E35.0-50-L23	1462	30609	219	23	10	129	5	111	1064	187
Nonlocal	92812-N42E35.0-50-L23	1461	32440	241	23	16	127	2	107	1032	177
Nonlocal	92815-N42E35.0-50-L23	1290	29055	220	24	14	125	5	108	1011	176
Nonlocal	92817-N42E35.0-50-L23	1438	29314	223	25	18	117	4	105	973	171
Nonlocal	92819-N42E35.0-50-L23	1554	32214	244	26	20	125	4	115	1069	181
Nonlocal	92827-N42E35-L23	1655	36184	265	23	19	149	4	131	1241	214
Nonlocal	92828-N42E35-L23	1014	30835	265	27	18	148	11	124	1147	198
Nonlocal	92830-N42E35.0-50-L23	1126	29820	261	23	19	133	14	100	999	184
Nonlocal	92831-N42E35.0-50-L23	1016	24298	229	21	16	119	4	104	988	175
Nonlocal	92832-N42E35.0-50-L23	1178	31504	191	22	18	127	8	114	1055	181
Nonlocal	92834-N42E35.0-50-L23	1543	35440	278	26	21	144	7	124	1168	202
Nonlocal	92847-N42E35.0-50-L23	1517	33226	259	25	18	129	4	116	1075	193
Nonlocal	92851-N42E35.0-50-L23	1508	34021	270	25	16	127	7	105	1055	180
Nonlocal	92863-N42E35.0-50-L23	1494	33068	240	23	20	142	7	119	1144	200

Table 3. Continued.

Nonlocal	92872-N42E35-L23	1209	31678	327	22	22	148	12	123	1224	210
Nonlocal	92876-N42E35-L23	1462	30427	244	26	16	145	6	119	1148	204
Nonlocal	92878-N42E35.o-50-L23	1500	33633	225	26	16	129	2	111	1045	184
Nonlocal	92879-N42E35.o-50-L23	1426	30098	240	25	19	145	4	127	1185	206
Nonlocal	92881-N42E35-L23	1371	31219	242	26	22	160	4	124	1257	231
Nonlocal	92883-N42E35.0-50-L23	1359	34426	248	22	18	137	7	126	1181	210
Unknown Group 3	91636-N42E35.0-50-L23 screened bag xrf#3	1734	41780	356	23	31	206	7	226	1728	338
Unknown Group 3	92576-N42E35.0-50-L23	1844	41582	358	22	34	225	8	237	1965	386
Unknown Group 3	92837-N42E35.0-50-L23	1740	43391	393	25	33	230	9	242	1928	365
Unknown Group 4	91616-N42E35.0-50-L23	2213	45220	535	23	49	292	9	345	2808	542
Unknown Group 4	92556-N42E35.0 -L22-xrf#12	2012	41145	503	23	39	259	7	316	2517	470
Unknown Group 4	92582-N42E35.0-50-L23	2305	45058	563	24	47	292	9	344	2789	514
Unknown Group 4	92824-N42E35-L23	2163	43765	521	24	42	285	8	322	2673	486
Unknown Group 5	91610-N42E35.0-50-L23	757	25985	119	22	14	121	9	66	595	124
Unknown Group 5	92556-N42E35.0 -L22-xrf#19	1043	21470	121	23	15	126	3	66	593	117
Unknown Group 6	91603-N42E35.0-50-L23	662	18681	156	22	8	69	5	92	647	90
Unknown Group 7	92814-N42E35-L23.	1013	24127	232	23	13	95	13	79	731	130

Here, we use the term "Nonlocal" for one of these geochemical groups, a cluster of 54 artifacts preserving relatively lower Y, Rb, and Nb (see Figure 7). Based on our understanding of the relationship between these elements in the MER, it is highly likely that the geographic source represented by this cluster is closer to the northern sources than to Baantu. Baantu has the highest Rb and Nb content of the known geographic sources, with Y on par with Ficke. Ficke is the next closest source that has been geochemically characterized using XRF (Shackley and Sahle 2017). These known relationships between trace elements would place the "Nonlocal" source somewhere around Bora, Worja, or

Alutu. The Rb content of the Nonlocal Group does overlap somewhat with the Bora ellipse, while the Nb content of a few samples overlaps with Alutu. More data is needed from the local SMER volcanic centers to examine these relationships.

Finally, comparing Baantu geochemical groups to MB1 samples analyzed prior to 2010 using gross peak counts indicates that the MB1 samples align with spectra peak counts and spectral patterns from the Baantu outcrop geochemical group (Figure 9; Table 4).

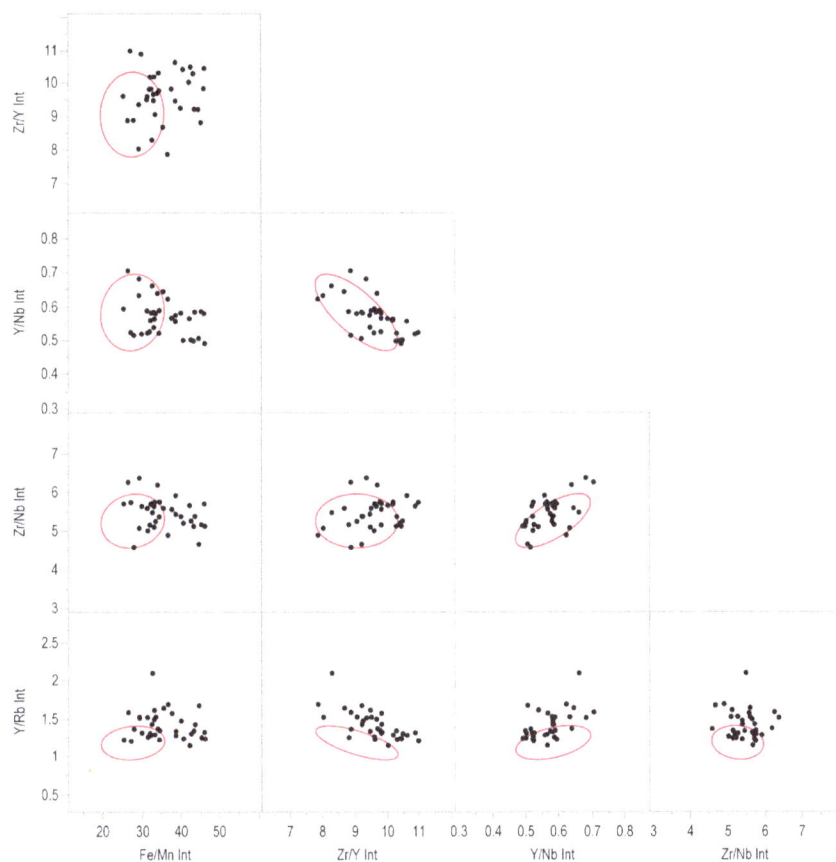

Figure 9. Peak intensity ratio matrix plot showing H9 (S-Group) artifacts. Baantu outcrop samples collected in 2020 form the red 95% confidence ellipse, and H9 artifacts analyzed by Warren (2010) are black dots. Figure by Smith and Johnson.

Table 4. Peak intensity and ratios for Baantu quarry outcrop samples reported here, and for the H9 (MB1) archaeological context reported in Warren (2010).

Specimen	Mn	Fe	Rb	Sr	Y	Zr	Nb	Fe/Mn	Zr/Y	Y/Nb	Zr/Nb	Y/Rb
BA_10	93	2861	294	45	338	3170	615	30.76	9.38	0.55	5.15	1.15
BA_11	78	1991	210	26	234	2207	418	25.53	9.43	0.56	5.28	1.11
BA_12	88	2470	265	42	282	2715	529	28.07	9.63	0.53	5.13	1.06
BA_13	101	2742	263	24	296	2901	550	27.15	9.80	0.54	5.27	1.13

Table 4. Continued.

BA_14	109	3183	303	35	356	3241	508	29.20	9.10	0.70	6.38	1.17
BA_15	102	3115	275	36	382	3253	623	30.54	8.52	0.61	5.22	1.39
BA_17	81	2508	259	38	292	2598	456	30.96	8.90	0.64	5.70	1.13
BA_18	106	2935	288	38	339	2971	579	27.69	8.76	0.59	5.13	1.18
BA_20	106	2769	263	31	321	2874	538	26.12	8.95	0.60	5.34	1.22
BA_21	106	2909	278	33	339	3020	563	27.44	8.91	0.60	5.36	1.22
BA_22	108	3276	316	46	392	3357	654	30.33	8.56	0.60	5.13	1.24
BA_23	111	3097	309	41	391	3257	607	27.90	8.33	0.64	5.37	1.27
BA_24	86	2241	239	36	292	2557	511	26.06	8.76	0.57	5.00	1.22
BA_25	110	3257	300	44	359	3292	665	29.61	9.17	0.54	4.95	1.20
BA_31	93	2738	262	40	304	2872	566	29.44	9.45	0.54	5.07	1.16
BA_32	99	2998	258	50	331	3077	586	30.28	9.30	0.56	5.25	1.28
BA_33	97	2854	280	52	339	2949	576	29.42	8.70	0.59	5.12	1.21
BA_35	74	2182	230	38	236	2400	413	29.49	10.17	0.57	5.81	1.03
BA_37	117	2992	302	29	348	3102	594	25.57	8.91	0.59	5.22	1.15
BA_4	108	2890	286	44	353	3125	628	26.76	8.85	0.56	4.98	1.23
BA_5	107	2944	252	30	326	3045	588	27.51	9.34	0.55	5.18	1.29
BA_6	96	2415	242	45	300	2575	515	25.16	8.58	0.58	5.00	1.24
BA_7	81	2476	244	32	253	2509	490	30.57	9.92	0.52	5.12	1.04
BA_8	105	2648	282	40	308	2706	494	25.22	8.79	0.62	5.48	1.09
BA_9	123	2901	318	46	325	3010	578	23.59	9.26	0.56	5.21	1.02
BA_19	99	3004	219	37	245	2454	492	30.34	10.02	0.50	4.99	1.12
BA_26	111	3334	320	41	369	3173	622	30.04	8.60	0.59	5.10	1.15
BA_27	107	2745	254	48	301	2839	555	25.65	9.43	0.54	5.12	1.19
BA_28	100	3439	313	32	410	3294	600	34.39	8.03	0.68	5.49	1.31
BA_30	173	2785	269	46	303	2762	523	16.10	9.12	0.58	5.28	1.13

Table 4. Continued.

BA_73	145	3281	222	42	304	2513	473	22.63	8.27	0.64	5.31	1.37
H9L37_BN9334	51	1610	198	13	250	2400	479	31.57	9.60	0.52	5.01	1.26
H9L37_BN9348	52	1725	149	10	223	2157	382	33.17	9.67	0.58	5.65	1.50
H9L37_BN9376	46	1735	141	11	222	2182	392	37.72	9.83	0.57	5.57	1.57
H9L37_BN9380	40	1627	196	19	243	2531	486	40.68	10.42	0.50	5.21	1.24
H9L37_BN9391	49	1800	150	10	254	1997	408	36.73	7.86	0.62	4.89	1.69
H9L37_BN9392	46	1289	150	8	205	1822	398	28.02	8.89	0.52	4.58	1.37
H9L37_BN9395	47	1819	163	12	209	2221	375	38.70	10.63	0.56	5.92	1.28
H9L37_BN9605	57	1712	195	14	257	2796	495	30.04	10.88	0.52	5.65	1.32
H9L37_BN9608	40	1175	122	11	186	1493	294	29.38	8.03	0.63	5.08	1.52
H9L37_BN9640	68	1728	181	6	222	2135	374	25.41	9.62	0.59	5.71	1.23
H9L38_BN10022	53	1712	168	9	216	2201	386	32.30	10.19	0.56	5.70	1.29
H9L38_BN10089	46	1499	172	10	246	2419	423	32.59	9.83	0.58	5.72	1.43
H9L38_BN11302	48	1859	210	12	281	2661	489	38.73	9.47	0.57	5.44	1.34
H9L38_BN11303	41	1897	219	15	271	2832	552	46.27	10.45	0.49	5.13	1.24
H9L38_BN11312	40	1379	120	8	147	1516	282	34.48	10.31	0.52	5.38	1.23
H9L38_BN11313	74	1965	160	12	254	2256	360	26.55	8.88	0.71	6.27	1.59
H9L38_BN11315	64	1744	157	12	190	2086	363	27.25	10.98	0.52	5.75	1.21
H9L38_BN11316	48	2048	213	13	278	2917	554	42.67	10.49	0.50	5.27	1.31
H9L38_BN11319	32	1397	124	4	177	1632	303	43.66	9.22	0.58	5.39	1.43
H9L38_BN12214	51	1500	140	8	212	1985	311	29.41	9.36	0.68	6.38	1.51
H9L38_BN13646	48	1599	157	13	203	2070	360	33.31	10.20	0.56	5.75	1.29
H9L38_BN9697	42	1774	182	7	209	2097	370	42.24	10.03	0.56	5.67	1.15
H9L39_BN10013	40	1844	175	10	232	2283	400	46.10	9.84	0.58	5.71	1.33
H9L39_BN10016	39	1768	193	17	242	2134	413	45.33	8.82	0.59	5.17	1.25
H9L39_BN10023	77	2522	256	20	538	4462	814	32.75	8.29	0.66	5.48	2.10

Table 4. Continued.

H9L39_BN9650	56	1991	205	12	337	2926	523	35.55	8.68	0.64	5.59	1.64
H9L39_BN9652	53	1667	136	7	207	1968	352	31.45	9.51	0.59	5.59	1.52
H9L39_BN9655	39	1742	179	10	300	2764	593	44.67	9.21	0.51	4.66	1.68
H9L39_BN9670	47	1602	156	5	214	2076	335	34.09	9.70	0.64	6.20	1.37
H9L39_BN9699	56	1858	138	11	223	2114	414	33.18	9.48	0.54	5.11	1.62
H9L45_16811	43	1863	206	7	277	2851	556	43.33	10.29	0.50	5.13	1.34
H9L45_BN15780	49	1694	155	11	209	2044	355	34.57	9.78	0.59	5.76	1.35
H9L45_BN15786	47	1883	194	8	286	2646	492	40.06	9.25	0.58	5.38	1.47
H9L45_BN15792	47	1578	131	4	200	1813	345	33.57	9.07	0.58	5.26	1.53
H9L45_BN15800	46	1476	180	16	236	2318	449	32.09	9.82	0.53	5.16	1.31

Discussion and conclusions

Negash and colleagues (2020) recently characterized 45 obsidian samples from sources across Ethiopia. Here, we demonstrate that studies of large source locations such as Baantu can benefit from extensive sampling to capture the range of possible geochemical variation within a source locality.

The large bifaces (Large Cutting Tools or "hand axes") and Levallois cores at Baantu give some indication of Baantu's deep history, but human engagement with this resource was likely also structured by regional geological history. Chernet, (2011: 134) cites an unpublished report of a K-Ar age of 1.57 from the inner wall of the "Obicha" caldera, but we do not know whether all Hobicha obsidian dates to the same period, or how many lava vents the caldera contains. Some have attempted to establish a tephrochronological framework for understanding landscape change within the Sodo region (WoldeGabriel et al. 1990; Bigazzi et al. 1993; Chernet 2011). If the last (and largest) regional volcanic event was the eruption of Mt. Duguna Fandgo ~450 ka BP (Bigazzi et al. 1993), quarrying and lithic manufacture at Baantu may have been interrupted by, or only begun after, this event. The deepest obsidian lenses and buried ashes at Baantu must be dated.

The "Unknown" chemical groups we identify in this study could represent local sub-sources, that is, as-yet unsampled outcrops within the Baantu area. They may also represent artifacts brought from other primary obsidian sources. Recycled artifacts (Figure 8) and grey/grey-green obsidian on the Baantu surface (grey-green obsidians have not been identified among Baantu outcrops) may be evidence that mobile groups were transporting, depositing, and reusing obsidian that was circulating within larger regional networks.

Future surveys in the Sodo region will identify the range of sources that would have been available to people living at Mochena Borago and other sites. We do not yet know where the MB5 Nonlocal Group artifacts came from, but it was the majority obsidian used prior to 50 ka BP at Mochena Borago. If these came from a more distant source than Baantu, this inverse distance decay relationship between source and site would have important implications for identifying direct procurement vs. intensive exchange (see Gould and Saggers 1985; Merrick and Brown 1984; Gamble 1993; Ambrose 2001; Ambrose 2012). These relationships can only be explored with more regional source data.

The extent to which the data we present here can be used to test the refugium hypotheses depends on the scale at which Late Pleistocene refugia are modeled. Fischer and colleagues (2021) have reconstructed altitudinal vegetation shifts in southernmost Ethiopia during the last glacial maximum (~25 ka BP), noting the expansion and contraction of moist highland habitats downslope during wet periods, as well as their contraction during arid periods. The Sodo region of the SMER currently lacks high resolution proxy data for vegetation change. However, it is likely that this area also witnessed altitudinal shifts in vegetation and food resources. Rivers that are today seasonal or diverted for agricultural purposes would have connected these biomes, and hunter-gatherers would have been able to move within and between these areas to exploit a variety of high, mid, and low-altitude environments (Vogelsang and Wendt 2018).

If during MIS 4 (~73.5–60 ka BP) and early MIS 3 (~60–28 ka BP; Hessler et al. 2010), hunter-gatherers responded to patchy resource distribution by broadening their foraging ranges, as was likely the case in Southern Africa (Ambrose and Lorenz 1990; McCall 2007), they would have encountered a variety of raw material sources dotted across the landscape. If these ranges contracted in response to stable resources in some areas, foraging procurement ranges might have contracted. The later situation may have been the context within which hunter-gatherers developed long-term material relations with the Baantu source. Regardless, we know that by Holocene times, Mochena Borago's occupants once again used a variety of obsidian types (see Negash 2022), as mobility and perhaps exchange once again increased.

This study contributes to a growing body of literature suggesting that Pleistocene hunter-gatherers in the Horn of Africa developed intensive economic ties to specific obsidian sources, which endured across major periods of social and ecological change (Negash et al. 2006, 2007; Ménard et al. 2014; Shackley and Sahle 2017; Hensel et al. 2021). Sites such as Mochena Borago Rockshelter containing well-dated, deep histories of lithic manufacture are a starting point from which to consider regional human engagements with stone landscapes through time. These engagements can only be explored through continued studies of hunter-gatherer toolstone procurement at the local and regional scale.

1 cm

Figure 8. A recycled biface/core from the Baantu surface.

Acknowledgments

The authors would like to thank Ato Girma Debusho and other members of the Wolaita Zonal Tourism offices for their invaluable support and knowledge. Thanks also to the curators, directors, and representatives at the Ethiopian Authority for Research and Conservation of Cultural Heritage (ARCCH). Thanks also to Ermias Yashitla, Olivia Kracht, and the Wolaita Sodo University students who assisted in surveying and collecting samples at Baantu. Funding was provided by the Department of Education Foreign Language Area Studies Program, the William J. Fulbright U.S. Student Program, the University of Florida Anthropology Department, the University of Florida International Center, and the University of Florida Center for African Studies. Thank you to the co-organizers of the 2021 International Obsidian Conference and co-editors of the current volume, Lucas R. M. Johnson, Kyle Freund, and Nico Tripcevich, for their invitation to contribute to this volume and their editorial assistance. Finally, thank you to M. Steven Shackley for his comments on this manuscript.

References

Ambrose, Stanly H.

2001   Paleolithic Technology and Human Evolution. *Science* 291 (5509): 1748–1753.

2012   Obsidian Dating and Source Exploitation Studies in Africa: implications for the evolution of human behavior. In *Obsidian and ancient manufactured glasses*, edited by Ioannis Liritzis and Christopher M. Stevenson, pp. 56–72. University of New Mexico Press, Albuquerque.

Ambrose, Stanley H., and Karl G. Lorenz

2022   1. Social and Ecological Models for the Middle Stone Age in Southern Africa. In *Emergence of Modern Humans: An Archaeological Perspective*, edited by Paul Mellars, pp. 3–33. Edinburgh University Press, Edinburgh.

Arthur, John W., Matthew C. Curtis, Kathryn J. W. Arthur, Mauro Coltorti, Pierluigi Pieruccini, Joséphine Lesur, Dorian Fuller, Leilani Lucas, Lawrence Conyers, Jay Stock, Sean Stretton, and Robert H. Tykot

2019   The Transition from Hunting-Gathering to Food Production in the Gamo Highlands of Southern Ethiopia. *African Archaeological Review* 36(1): 5–65.

Benito-Calvo, Alfonso, Ignacio de la Torre, Rafael Mora, Dawit Tibebu, Noemi Morán, and Jorge Martínez-Moreno

2007   Geoarchaeological Potential of the Western Bank of the Bilate River (Ethiopia). *Analele Universitatii din Oradea XVIISeria Geographie*: 99–104.

Beyin, Amanuel

2006   The Bab al Mandab vs. the Nile-Levant: An Appraisal of the Two Dispersal Routes for Early Modern Humans Out of Africa. *African Archaeological Review* 23(1): 5–30.

Bigazzi, Giulio., F. P. Bonadonna, G. M. Di Paola, and A. Giuliani

1993   K-Ar and Fission Track Ages of the Last Volcano-tectonic Phase in the Ethiopian Rift Valley (Tullu Moye' Area). In *Geology and Mineral Resources of Somalia and Surrounding Regions*, edited by E. Abbate, M. Sagri, and F. P. Sassi, pp. 311–322. Oltremare: Florence, Italy.

Blegen, Nick

2017   The Earliest Long-Distance Obsidian Transport: Evidence from the ⬚200 ka Middle Stone Age Sibilo School Road Site, Baringo, Kenya. *Journal of Human Evolution* 103: 1–19.

Brandt, Steven A.

1988   Early Holocene Mortuary Practices and Hunter-Gatherer Adaptations in Southern Somalia. *World Archaeology* 20(1): 40–56.

Brandt, Steven A., Elisabeth A. Hildebrand, Ralf Vogelsang, Erich C. Fisher, Brady Kelsey, Peter Lanzarone, Hannah Parow-Souchon, Benjamin D. Smith, Abebe Mengistu Teffera, Joséphine Lesur, and Kylie Bermensolo
    2023   Mochena Borago Rockshelter, SW Ethiopian Highlands. In *Handbook of Pleistocene Archaeology of Africa: Hominin Behavior, Geography, and Chronology*, edited by Amanuel Beyin, David K. Wright, Jayne Wilkins, and Deborah I. Olsezewski, pp. 461–482. Springer Cham.

Brandt, Steven, Elisabeth Hildebrand, Ralf Vogelsang, Jesse Wolfhagen, and Hong Wang
    2017   A New MIS 3 Radiocarbon Chronology for Mochena Borago Rockshelter, SW Ethiopia: Implications for the Interpretation of Late Pleistocene Chronostratigraphy and Human Behavior. *Journal of Archaeological Science: Reports* 11: 352–369.

Brandt, Steven A., Erich C. Fisher, Elisabeth A. Hildebrand, Ralf Vogelsang, Stanley H. Ambrose, Joséphine Lesur, and Hong Wang
    2012   Early MIS 3 Occupation of Mochena Borago Rockshelter, Southwest Ethiopian Highlands: Implications for Late Pleistocene Archaeology, Paleoenvironments and Modern Human Dispersals. *Quaternary International* 274: 38–54.

Brooks, Alison S., John E. Yellen, Richard Potts, Anna K. Behrensmeyer, Alan L. Deino, David E. Leslie, Stanley H. Ambrose, Jeffrey R. Ferguson, Francesco d'Errico, Andrew M. Zipkin, Scott Whittaker, Jeffrey Post, Elizabeth G. Veatch, Kimberly Foecke, and Jennifer B. Clark
    2018   Long-Distance Stone Transport and Pigment Use in the Earliest Middle Stone Age. *Science* 360(6384): 90–94.

Brown, F. H., B. P. Nash, D. P. Fernandez, H. V. Merrick, and R. J. Thomas
    2013   Geochemical Composition of Source Obsidians from Kenya. *Journal of Archaeological Science* 40(8): 3233–3251.

Cann, Johnson Robin, and Colin Renfrew
    1964   The Characterization of Obsidian and Its Application to the Mediterranean Region. *Proceedings of the Prehistoric Society* 30: 111–133.

Chernet, Tadiwos
    2011   Geology and Hydrothermal Resources in the Northern Lake Abaya Area (Ethiopia). *Journal of African Earth Sciences* 61(2): 129–141.

Clark, J. Desmond
    1988   The Middle Stone Age of East Africa and the Beginnings of Regional Identity. *Journal of World Prehistory* 2(3): 235–305.

Clark, J. Desmond, Kenneth D. Williamson, Joseph W. Michels, and Curtis A. Marean
1984   A Middle Stone Age Occupation Site at Porc Epic Cave, Dire Dawa (East-Central Ethiopia). *The African Archaeological Review* 2(1): 37–71.

de la Torre, Ignacio, Alfonso Benito-Calvo, Rafael Mora, Jorge Martínez-Moreno, Noemi Morán, and Dawit Tibebu
2007   Stone Age Occurrences in the Western Bank of the Bilate River (Southern Ethiopia) – Some Preliminary Results. *Nyame Akuma* (67): 14–25.

Ferguson, Jeffrey R
2012   X-Ray Fluorescence of Obsidian: Approaches to Calibration and the Analysis of Small Samples. In *Handheld XRF for art and archaeology*, vol. 2, edited by Aaron N. Shugar and Jennifer L. Mass, pp. 401–422. Leuven University Press, Leuven.

Fischer, Markus L., Felix Bachofer, Chad L. Yost, Ines J. E. Bludau, Christian Schepers, Verena Foerster, Henry Lamb, Frank Schäbitz, Asfawossen Asrat, Martin H. Trauth, and Annett Junginger
2021   A Phytolith Supported Biosphere-Hydrosphere Predictive Model for Southern Ethiopia: Insights into Paleoenvironmental Changes and Human Landscape Preferences since the Last Glacial Maximum. *Geosciences* 11(10): 418.

Frahm, Ellery
2013a   Validity of "Off-the-shelf" Handheld Portable XRF for Sourcing Near Eastern Obsidian Chip Debris. *Journal of Archaeological Science* 40(2): 1080–1092.
2013b   Is Obsidian Sourcing about Geochemistry or Archaeology? A Reply to Speakman and Shackley. *Journal of Archaeological Science* 40(2): 1444–1448.
2019   Introducing the Peabody-Yale Reference Obsidians (PYRO) Sets: Open-source Calibration and Evaluation Standards for Quantitative X-ray Fluorescence Analysis. *Journal of Archaeological Science: Reports* 27: 101957.

Frahm, Ellery, Steven T. Goldstein, and Christian A. Tryon
2017   Late Holocene Forager-Fisher and Pastoralist Interactions along the Lake Victoria Shores, Kenya: Perspectives from Portable XRF of Obsidian Artifacts. *Journal of Archaeological Science: Reports* 11: 717–742.

Francaviglia, Vincenzo M.
1990   Les gisements d'obsidienne hyperalcaline dans l'ancien monde : étude comparative. *Revue d'Archéométrie* 14(1): 43–64.

Giménez, Javier, Josep A. Sánchez, and Lluís Solano
2015   Identifying the Ethiopian Origin of the Obsidian Found in Upper Egypt (Naqada period) and the Most Likely Exchange Routes. *The Journal of Egyptian Archaeology* 101(1): 349–359.

Glascock, Michael D., Amanuel Beyin, and Magen E. Coleman
2008   X-Ray Fluorescence and Neutron Activation Analysis of Obsidian from the Red Sea Coast of Eritrea. International Association for Obsidian Studies. *IAOS Bulletin* 2008: 6–11.

Gould, Richard A., and Sherry Saggers
1985   Lithic Procurement in Central Australia: A Closer Look at Binford's Idea of Embeddedness in Archaeology. *American Antiquity* 50(1): 117–136.

Gutherz, Xavier, L. Jallot, R. Joussaume, and B. Poisblaud
1998   Abri sous-roche de Moche Borago (Wolayta-Soddo). Campagne de sondages, novembre unpublished report.

Gutherz, Xavier
2000   Sondages dans l'abri sous-roche de Moche Borago Gongolo dans le Wolayta (Éthiopie). *Annales d'Éthiopie* 16(1): 35–38.

Hensel, Elena A., Ralf Vogelsang, Tom Noack, and Olaf Bubenzer
2021   Stratigraphy and Chronology of Sodicho Rockshelter – A New Sedimentological Record of Past Environmental Changes and Human Settlement Phases in Southwestern Ethiopia. *Frontiers in Earth Science* 8. https://doi.org/10.3389/feart.2020.611700 (accessed January 21, 2022.)

Hessler, Ines, Lydie Dupont, Raymonde Bonnefille, Hermann Behling, Catalina González, Karin F. Helmens, Henry Hooghiemstra, Judicael Lebamba, Marie-Pierre Ledru, Anne-Marie Lézine, Jean Maley, Fabienne Marret, and Annie Vincens
2010   Millennial-scale Changes in Vegetation Records from Tropical Africa and South America during the Last Glacial. *Quaternary Science Reviews* 29(21): 2882–2899.

Hillis, A., J. Feinberg, E. Frahm, and C. Johnson
2010   Magnetic Sourcing of Obsidian Artifacts: Successes and Limitations. American Geophysical Union, Fall Meeting 2010, abstract id. GP43A-1041

Johnson, Lucas R. Martindale, Jeffrey R. Ferguson, Kyle P. Freund, Lee Drake, and Daron Duke
2021   Evaluating Obsidian Calibration Sets with Portable X-Ray Fluorescence (ED-XRF) Instruments. *Journal of Archaeological Science: Reports* 39: 103126.

Kazmin, V., and Kaz'min Vg
1979   Transform Vaults in the East African Rift System. In *International symposium: Geodynamic Evolution of the Afro-Arabian Rift System, Abstracts*, pp. 54–56. Accademia Nazionale dei Lincei, Rome, Italy.

Khalidi, Lamya, Carlo Mologni, Clément Ménard, Lucie Coudert, Marzia Gabriele, Gourguen Davtian, Jessie Cauliez, Joséphine Lesur, Laurent Bruxelles, Lorène Chesnaux, Blade Engda Redae, Emily Hainsworth, Cécile Doubre, Marie Revel, Mathieu Schuster, and Antoine Zazzo
    2020   9,000 Years of Human Lakeside Adaptation in the Ethiopian Afar: Fisher-Foragers and the First Pastoralists in the Lake Abhe Basin during the African Humid Period. *Quaternary Science Reviews* 243: 106459.

Kurashina, Hiroyasu
    1978   An Examination of Prehistoric Lithic Technology in East-Central Ethiopia. Unpublished PhD dissertation, University of California, Berkeley, California.

Lesur, Joséphine, Jean-Denis Vigne, and Xavier Gutherz
    2007   Exploitation of Wild Mammals in South-west Ethiopia during the Holocene (4000 BC–500 AD): The Finds from Moche Borago Shelter (Wolayta). *Environmental Archaeology* 12 (2): 139–159.

McBrearty, Sally, and Alison S. Brooks
    2000   The Revolution that Wasn't: A New Interpretation of the Origin of Modern Human Behavior. *Journal of Human Evolution* 39(5): 453–563.

McCall, Grant S.
    2007   Behavioral Ecological Models of Lithic Technological Change during the Later Middle Stone Age of South Africa. *Journal of Archaeological Science* 34(10): 1738–1751.

McDougall, J. M., D. H. Tarling, and S. E. Warren
    1983   The Magnetic Sourcing of Obsidian Samples from Mediterranean and Near Eastern Sources. *Journal of Archaeological Science* 10(5): 441–452.

Ménard, Clément, François Bon, Asamerew Dessie, Laurent Bruxelles, Katja Douze, François-Xavier Fauvelle, Lamya Khalidi, Joséphine Lesur, and Romain Mensan
    2014   Late Stone Age Variability in the Main Ethiopian Rift: New Data from the Bulbula River, Ziway-Shala Basin. *Quaternary International* 343: 53–68.

Merrick, Harry V., and Francis H. Brown
    1984   Obsidian Sources and Patterns of Source Utilization in Kenya and Northern Tanzania: Some Initial Findings. *The African Archaeological Review* 2(1): 129–152.

Muir, I. D., and F. Hivernel
    1976   Obsidians from the Melka-Konture Prehistoric Site, Ethiopia. *Journal of Archaeological Science* 3(3): 211–217.

Negash, Agazi

2022   Geological Origin of the Archaeological Obsidian Artifacts from the Holocene Sequences of Mochena Borago and Yabello, Southern Ethiopia. *Archaeometry*: arcm.12772.

Negash, Agazi, Mulugeta Alene, F. H. Brown, B. P. Nash, and M. S. Shackley

2007   Geochemical Sources for the Terminal Pleistocene/early Holocene Obsidian Artifacts of the Site of Beseka, Central Ethiopia. *Journal of Archaeological Science* 34(8): 1205–1210.

Negash, Agazi, F. H. Brown, Mulugeta Alene, and B. Nash

2010   Provenance of Middle Stone Age Obsidian Artefacts from the Central Sector of the Main Ethiopian Rift Valley. *SINET: Ethiopian Journal of Science* 33(1): 21–30.

Negash, Agazi, F. Brown, and B. Nash

2011   Varieties and Sources of Artefactual Obsidian in the Middle Stone Age of the Middle Awash, Ethiopia. *Archaeometry* 53(4): 661–673.

Negash, Agazi, Barbara P. Nash, and Francis H. Brown

2020   An Initial Survey of the Composition of Ethiopian Obsidian. *Journal of African Earth Sciences* 172: 103977.

Negash, Agazi, and M. S. Shackley

2006   Geochemical Provenience of Obsidian Artifacts from the MSA Site of Porc Epic, Ethiopia. *Archaeometry* 48(1): 1–12.

Negash, Agazi, M. Steven Shackley, and Mulugeta Alene

2006   Source Provenance of Obsidian Artifacts from the Early Stone Age (ESA) Site of Melka Konture, Ethiopia. *Journal of Archaeological Science* 33(12): 1647–1650.

Oppenheimer, Clive, Lamya Khalidi, Bernard Gratuze, Nels Iverson, Christine Lane, Céline Vidal, Yonatan Sahle, Nick Blegen, Ermias Yohannes, Amy Donovan, Berhe Goitom, James O. S. Hammond, Edward Keall, Ghebrebrhan Ogubazghi, Bill McIntosh, and Ulf Büntgen

2019   Risk and Reward: Explosive Eruptions and Obsidian Lithic Resource at Nabro Volcano (Eritrea). *Quaternary Science Reviews* 226: 105995.

Piperno, Marcello, Carmine Collina, Rosalia Gallotti, Jean-Paul Raynal, Guy Kieffer, Francois-Xavier le Bourdonnec, Gerard Poupeau, and Denis Geraads

2009   Obsidian Exploitation and Utilization during the Oldowan at Melka Kunture (Ethiopia). In *Interdisciplinary Approaches to the Oldowan*, edited by Erella Hovers and David R. Braun, pp. 111–128. Vertebrate Paleobiology and Paleoanthropology. Springer, Dordrecht, Netherlands.

Rose, Jeffrey I., Vitaly I. Usik, Anthony E. Marks, Yamandu H. Hilbert, Christopher S. Galletti, Ash Parton, Jean Marie Geiling, Viktor Černý, Mike W. Morley, and Richard G. Roberts

2011   The Nubian complex of Dhofar, Oman: An African middle stone age industry in Southern Arabia. *PLoS ONE* 6(11).

Sahle, Yonatan, W. Karl Hutchings, David R. Braun, Judith C. Sealy, Leah E. Morgan, Agazi Negash, and Balemwal Atnafu

2013   Earliest Stone-tipped Projectiles from the Ethiopian Rift Date to > 279,000 Years Ago. *PLoS one* 8(11): e78092.

Schepers, Christian

2019   By the Rivers of Bilate – Twisted Bifaces from Ethiopia. In *61st Annual Meeting of the Hugo Obermaier-Society: New Perspectives on Neanderthal Behaviour*, 23. April 27, 2019, Erkrath. Poster.

Shackley, M. Steven (editor)

2011   *X-Ray Fluorescence Spectrometry (XRF) in Geoarchaeology.* Springer, New York.

Shackley, M. Steven, and Yonatan Sahle

2017   Geochemical Characterization of Four Quaternary Obsidian Sources and Provenance of Obsidian Artifacts from the Middle Stone Age Site of Gademotta, Main Ethiopian Rift. *Geoarchaeology* 32(2): 302–310.

Speakman, Robert J., and M. Steven Shackley

2013   Silo science and portable XRF in Archaeology: A Response to Frahm. *Journal of Archaeological Science* 40(2): 1435–1443.

Stewart, Brian A., and Sacha C. Jones

2016   Africa from MIS 6-2: The Fluorescence of Modern Humans. In *Africa from MIS 6-2.* Edited by S. Jones and B. Stewart. Vertebrate Paleobiology and Paleoanthropology. Springer, Dordrecht, Netherlands.

van Peer, Philip

1992   *The Levallois Reduction Strategy.* Prehistory Press, Wisconsin.

Vogel, Nadia, Sebastien Nomade, Agazi Negash, and Paul R. Renne

2006   Forensic ^{40}Ar/^{39}Ar Dating: A Provenance Study of Middle Stone Age Obsidian Artifacts from Ethiopia. *Journal of Archaeological Science* 33(12): 1749–1765.

Vogelsang, Ralf, and Karl Peter Wendt

    2018   Reconstructing Prehistoric Settlement Models and Land Use Patterns on Mt. Damota/SW Ethiopia. *Quaternary International* 485. Inside – Outside: Integrating Cave and Open-Air Archives: 140–149.

Warren, Andrea M.

    2010   Preliminary Characterization and Provenience of Obsidian Artifacts from Ethiopian Archaeological Sites Using Portable X-Ray Fluorescence. Unpublished BA thesis, University of Florida, Gainesville.

WoldeGabriel, Giday, James L. Aronson, and Robert C. Walter

    1990   Geology, Geochronology, and Rift Basin Development in the Central Sector of the Main Ethiopia Rift. *Geological Society of America Bulletin* 102(4): 439–458.

Yamazaki, Dai, Daiki Ikeshima, Ryunosuke Tawatari, Tomohiro Yamaguchi, Fiachra O'Loughlin, Jeffery C. Neal, Christopher C. Sampson, Shinjiro Kanae, and Paul D. Bates

    2017   A high-accuracy map of global terrain elevations. *Geophysical Research Letters* 44(11): 5844–5853.

CHAPTER 6

XRF Provenance Analysis of the Obsidian Jewelry from the Great Temple of Tenochtitlan, Mexico

EMILIANO MELGAR TÍSOC, GUILLERMO ACOSTA-OCHOA, VÍCTOR GARCÍA-GÓMEZ, REYNA SOLÍS-CIRIACO, LUIS COBA-MORALES, AND EDER BORJA-LAGUNA

Emiliano Melgar Tísoc Templo Mayor Museum, Seminario 8, Col. Centro, Cuauhtemoc, Mexico City, CDMX, C.P. 06060, Mexico (emiliano_melgar@inah.gob.mx, corresponding author)

Guillermo Acosta-Ochoa Instituto de Investigaciones Antropológicas, CU-UNAM, Coyoacan, Mexico City, CDMX, C.P. 04510, Mexico

Víctor García-Gómez Instituto de Investigaciones Antropológicas, CU-UNAM, Coyoacan, Mexico City, CDMX, C.P. 04510, Mexico

Reyna Solís-Ciriaco Templo Mayor Museum, Seminario 8, Col. Centro, Cuauhtemoc, Mexico City, CDMX, C.P. 06060, Mexico

Luis Coba-Morales Instituto de Investigaciones Antropológicas, CU-UNAM, Coyoacan, Mexico City, CDMX, C.P. 04510, Mexico

Eder Borja-Laguna Instituto de Investigaciones Antropológicas, CU-UNAM, Coyoacan, Mexico City, CDMX, C.P. 04510, Mexico

Abstract

In the Great Temple of Tenochtitlan, archaeologists have recovered more than 1,000 artifacts of chipped obsidian, but only 210 pieces of jewelry. Surprisingly, during 43 years of excavations, only one provenance study with Neutron Activation Analysis (NAA) had been carried out on seven objects (five blades, one flake, and one pendant). In this work, we will present the results of recent analysis with portable X-ray fluorescence (pXRF) applied on 19 obsidian lapidary goods with different morphology, function, temporality, and color. By analyzing the principal components (PCA) of the obsidian artifacts from the Great Temple, compared to geological samples, this statistical study allowed us to identify 5 of the 10 elements (Y, Rb, Sr, Zr, and Nb) that are the most significant to defining provenance. Based on that, we could determine that most of the samples belong to the main deposits of the Basin of Mexico (Otumba and Pachuca), which is a common sourcing pattern among the assemblages of central Mexico during the Postclassic period (AD 1200–1521). However, two distinct deposits could be defined better by Hierarchical Cluster Analysis. Thus, three objects came from Pico de Orizaba in Veracruz and one from Ucareo in Michoacan. The interest of the Mexica in the Pico de Orizaba source was important for the military garrisons located about 20 km from the mines. In contrast, the Ucareo material was an unexpected result because this source was unusual in the Basin of Mexico, most likely due to the dominance of the mine by the Tarascan Empire in West Mexico, one of the principal rivals of Tenochtitlan. This object could be a relic, obtained by looting ancient sites of the Basin of Mexico with prior occupation or as a gift or war prize. Finally, it would be the first reported material originating in that region in the Great Temple assemblage.

Introduction

Since the first study of the recovered objects from the excavations of the Templo Mayor Project (Great Temple of Tenochtitlan) of 1978, efforts have been made to establish a chronological comparison between the various offerings, as well as the similarities or differences between their contents, since it has been suggested that, with each new ruler that ascended to the throne, there was the need to surpass the achievements of his predecessor, consolidate his power by dominating new territories, and make more sumptuous offerings and ceremonies (Vilanova de Allende 2002: 100). This becomes clear when comparing the offerings of later stages such as Stage IVb, assigned to Axayácatl (AD 1469–1481), with previous stages (Matos Moctezuma 1989: 119); indeed, the later stage is found to demonstrate the most variety and wealth (Olmo Frese 1999: 65). But what about the periods of expansionist boom, such as the reign of Ahuítzotl that goes from AD 1486 to 1502 of Stage VI, or the periods of military crises, such as those of Axayácatl and Tízoc (Obregón Rodríguez 1995: 287)? How are they reflected in the content of the offerings?

Despite these limitations and the differences in materials between construction stages, Matos Moctezuma (1988: 88) suggests that most of the foreign, or non-Aztec, materials found in the offerings came from distant areas or provinces that were subject to Mexica rule and that none came from independent areas such as the Tarascan and Maya. However, recent provenance and manufacturing studies have confirmed that jadeite objects with Mayan technology are present in the offerings of the Great Temple (Melgar Tísoc et al. 2018; Monterrosa Desruelles 2018). Therefore, it is necessary to perform laboratory analyses to identify the raw materials from which the objects were made, in addition to the techno-stylistic study of the objects. This gives a fuller picture of the economic matter: further information regarding the provenance of raw materials, their manufacture, and circulation should indicate their ethnic affiliation (Melgar Tísoc 2014; Melgar Tísoc et al. 2018).

Because most of the obsidian lapidary objects recovered from the offerings of the Great Temple are finished artifacts—such as earplugs, duck-headed pendants, and lip plugs—there has been a search for similarities with pieces from other contemporary sites, such as Tlatelolco, Texcoco, and Xochimilco (Figure 1), to determine their foreign origin.

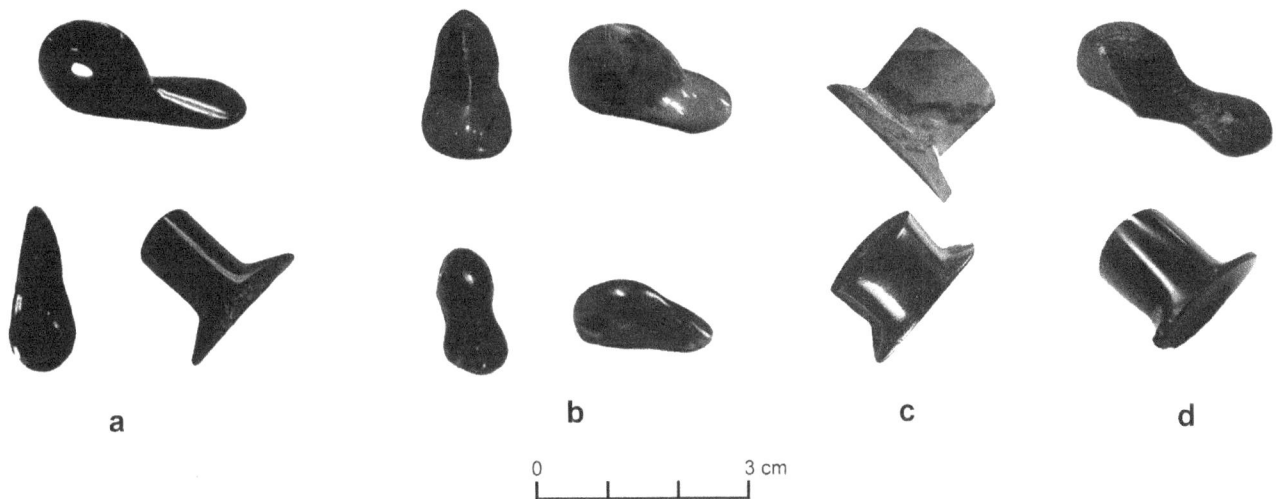

Figure 1. Obsidian lapidary objects from (a) Tenochtitlan, (b) Tlatelolco, (c) Texcoco, and (d) Xochimilco. Photos by Emiliano Melgar Tísoc and Reyna Solís Ciriaco.

However, some similar objects recovered from Tenochtitlan were found in pre-Mexica settlements, such as the open-work discs from Tula. In contrast, others are unique pieces, such as the funerary urn decorated with a skull. Based on the above, Matos Moctezuma (1988: 11) has suggested that these goods belong to an "ancient tradition" of carving and polishing obsidian from central Mexico. Likewise, there are pieces of standardized forms—such as scepters, ear flares, and nose plugs of Nahua divinities—that are considered to be Mexica or Aztec artifacts (Matos Moctezuma 1988: 92; López Luján 1993: 138–139). Different colors are also present in the obsidian objects, predominantly the green-golden obsidian from Sierra de Pachuca (Athié Islas 2001: 59), but there are also gray and reddish obsidians (Athié Islas 2001: 63). For all these reasons, it is important to determine the provenance and manufacture of the pieces to investigate the different ways of acquisition, circulation, and transformation of these precious goods. In the case of the manufacturing techniques, there are several studies on the characterization of the work traces of the pieces (Melgar Tísoc and Solís Ciriaco 2009; Velázquez Castro and Melgar Tísoc 2014; Solís Ciriaco 2018). In contrast, there is only one reported analysis on the provenance of the obsidian objects found in the offerings of the Great Temple of Tenochtitlan (Athié Islas 2001). Considering this, the present research has focused on analyzing obsidians of different shades to identify their deposits of origin through their geochemical characterization using pXRF equipment.

Previous studies on the provenance of obsidian in the offerings of the Great Temple

The obsidian artifacts from the Great Temple have practically no studies focused on their provenance. The only work performed was by NAA, which was applied to six samples (five blades and one flake) and identified that the green and green-golden obsidian came from Pachuca (also known as Sierra de las Navajas), one reddish and two grays from Otumba, and the last grayish one from Zaragoza (Glascock and Neff 1999; Athié Islas 2001: 61–63).

Unfortunately, there were no geochemical characterization analyses performed on the obsidian lapidary from the Great Temple of Tenochtitlan. Because of this, the provenance had been determined by its macroscopic characteristics. The most common attribute is its coloration, and based on it, three colors were identified: Almost all lapidary objects that were made of green and green-golden obsidian come from Sierra de Pachuca (Athié Islas 2001: 114–130, 134–137). Only three types of objects (helical beads, pumpkin beads, and globular beads) were made on red-orange obsidian with a black background, as well as on gray obsidian—these are considered to be from Otumba (Athié Islas 2001: 130–133).

Due to issues of color identification, the possibility that some gray obsidian objects come from other sourcing areas, and the fact that grayish obsidians share optical properties among various sites (e.g., Ucareo and Zaragoza), visual identification is not enough to confirm specific deposits (Braswell 1997). It should be noted that the reddish pieces are not exclusive to the Otumba site, as they occasionally appear in other sourcing areas from central Mexico and are common in deposits located in Michoacán and Jalisco (Cobean 2002). Therefore, the need to carry out chemical analysis with the pXRF equipment was clear to us.

Analyzing the obsidian lapidary from the Great Temple

We employed pXRF equipment to determine the provenance of the obsidian lapidary objects from the offerings of the Great Temple of Tenochtitlan. A representative sample was selected considering the diversity of colors (mainly different gray and reddish pieces), contexts, and constructive stages.

Based on this, 19 pieces were chosen from 4 offerings (Table 1), comprising 8 types of objects: urn, urn lid, helical bead, pumpkin bead, globular bead, duck head pendant, small cylindrical scepter with globular top, and giant cylindrical scepter with globular top. These artifacts came from 3 offerings inside the Great Temple (known as *Huey Teocalli*); and the last offering came from the House of the Eagles (known as Casa de las Águilas) to be compared with structures of the Sacred Precinct of Mexico Tenochtitlan (Figures 2–4).

Table 1. Obsidian objects analyzed from the Great Temple of Tenochtitlan.

Stage	Object	Quantity	Offering	Structure	Ruler	Color
II	Funerary urn	1	34	Great Temple	Acamapichtli, Huizilíhuitl and Chimalpopoca	Green
	Urn lid	1				Green
	Globular bead	1				Reddish brown
	Helical beads	2	39			Gray
		1				Green
	Pumpkin beads	3				Reddish brown
IVb	Duck head pendants	5	14		Axayácatl	Green
V	Small cylindrical scepter with globular top	1	V	House of the Eagles	Tízoc	Golden
	Duck head pendants	3				Gray
	Broken giant cylindrical scepter with globular top	2				Golden

Figure 2. Examples of the obsidian pieces analyzed. Photos by Emiliano Melgar Tísoc.

Figure 3. The offerings with obsidian pieces analyzed. Drawing by Emiliano Melgar Tísoc and Víctor Solís Ciriaco.

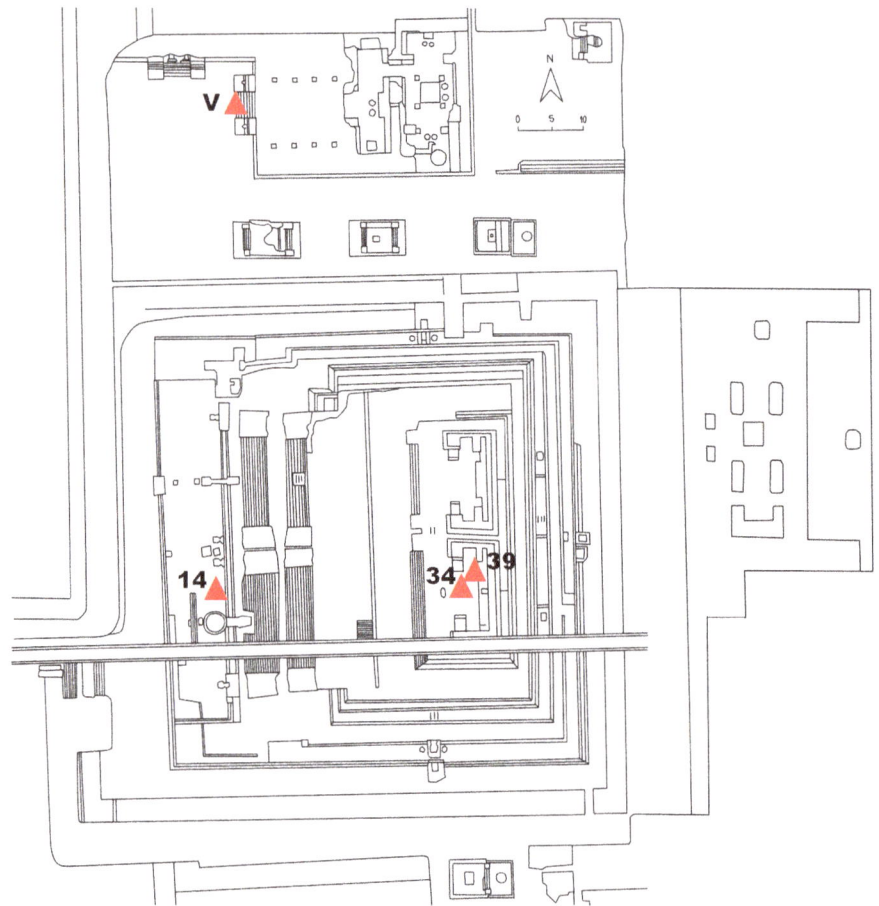

Most of the gray and reddish obsidian objects are concentrated in two different offerings (Offerings 34 and 39) of Stage II (AD 1375–1426), corresponding to three Mexica rulers (Acamapichtli, Huitzilíhuitl, and Chimalpopoca) who ruled most of the pre-imperial period of Tenochtitlan. In the latter stages, practically all the obsidian lapidary items were crafted on green-golden obsidian from Sierra de Pachuca—the only exception is the transparent gray duck head pendants from Offering V of constructive Stage V (AD 1481–1486), corresponding to the government of Tízoc. Interestingly, and for comparison purposes, these grayish objects break the uniformity of green-golden pieces made with obsidian from Sierra de Pachuca of the imperial stages (Melgar Tísoc and Solís Ciriaco 2009).

XRF analysis methodology

X-ray fluorescence (XRF) analysis is particularly useful because it is a reliable, nondestructive, and relatively inexpensive elemental analysis that is being used increasingly in archaeology since the pioneering study of Robert Jack and Robert F. Heizer in 1968. Unlike other techniques (such as NAA), XRF analysis operates directly on the object and with minimal preparation of the samples.

This analytical technique is based on the interaction between A) the electrons of the different atoms that make up the objects to be analyzed and B) primary X-ray photons coming from the analytical equipment. When the atoms are excited with high short-wave radiation (X-rays), their inner orbital electrons dislocate, causing their instability until another electron from an outer orbital replaces the vacant space left by the first electron. Because of the energy differences between the electrons in the inner (lower binding energy) and outer (higher binding energy) orbitals, the result of this orbital shift process is the release of radiation of lower energy than the primary incident rays in the form of fluorescence (Shackley 2011: 28). Since the energy emitted is unique to each chemical element, we can quantify its relative abundance using certified standards of known composition, allowing the chemical element that emitted the radiation to be identified.

Obsidian is an archaeological material that is ideal for analysis using the XRF technique (Glascock 2011; Bonsall et al., this volume) due to its continuous use by prehispanic societies, its degree of preservation, and the fact that it is a homogeneous material with a specific composition of trace elements for each geological event that gave rise to the different exploited deposits. With our study, only a minimal cleaning of the surface to be analyzed was done; further, we ensured that the surfaces were flat or slightly convex, covered the detector area (8 mm), and complied with the minimum thickness of 0.35 cm, according to the methodology established by the Laboratory of Prehistory and Evolution (LAPE) at the National Autonomous University of Mexico (UNAM) (Acosta et al. 2015).

For this study, we evaluated the concentrations, in parts per million (ppm), of 10 of the most representative chemical elements in archaeological and geological samples (Mn, Fe, Zn, Ga, Th, Rb, Sr, Y, Zr, Nb). The analyses were performed in the LAPE using a Bruker portable equipment (Tracer III-V SD) configured at 40keV, 25 µA, with a 12 mil Al, 1 mil Ti, and 6 mil Cu filter. Samples were irradiated for 200 seconds, and we quantified the resulting spectra to convert the results to ppm, employing the empirical calibration developed by Speakman (2012). This calibration allows us to obtain comparable results with other methods such as NAA and has been validated by previous archaeological studies (Nazaroff et al. 2010; Reuther et al. 2011). Finally, to evaluate the precision and accuracy of the equipment, readings of the SRM278 (Obsidian Rock) standard were taken at the beginning and end of each three-hour work session.

To determine the provenance of the archaeological obsidians, the results were compared with the database of geological samples from 15 deposits sampled by the LAPE (García Gómez 2018). The geological samples were gathered from deposits from the Trans-Mexican Neovolcanic Belt and the Guatemalan Highlands. However, based on the results in our study, only samples from the former region were used to illustrate the comparison among archaeological and geological pieces (Figures 4 and 5).

The values obtained with the compositional analysis (Table 2) were statistically processed using the software *Past*® to determine the provenance, based on the similarity of the chemical fingerprint of the archaeological specimens compared to the geological samples.

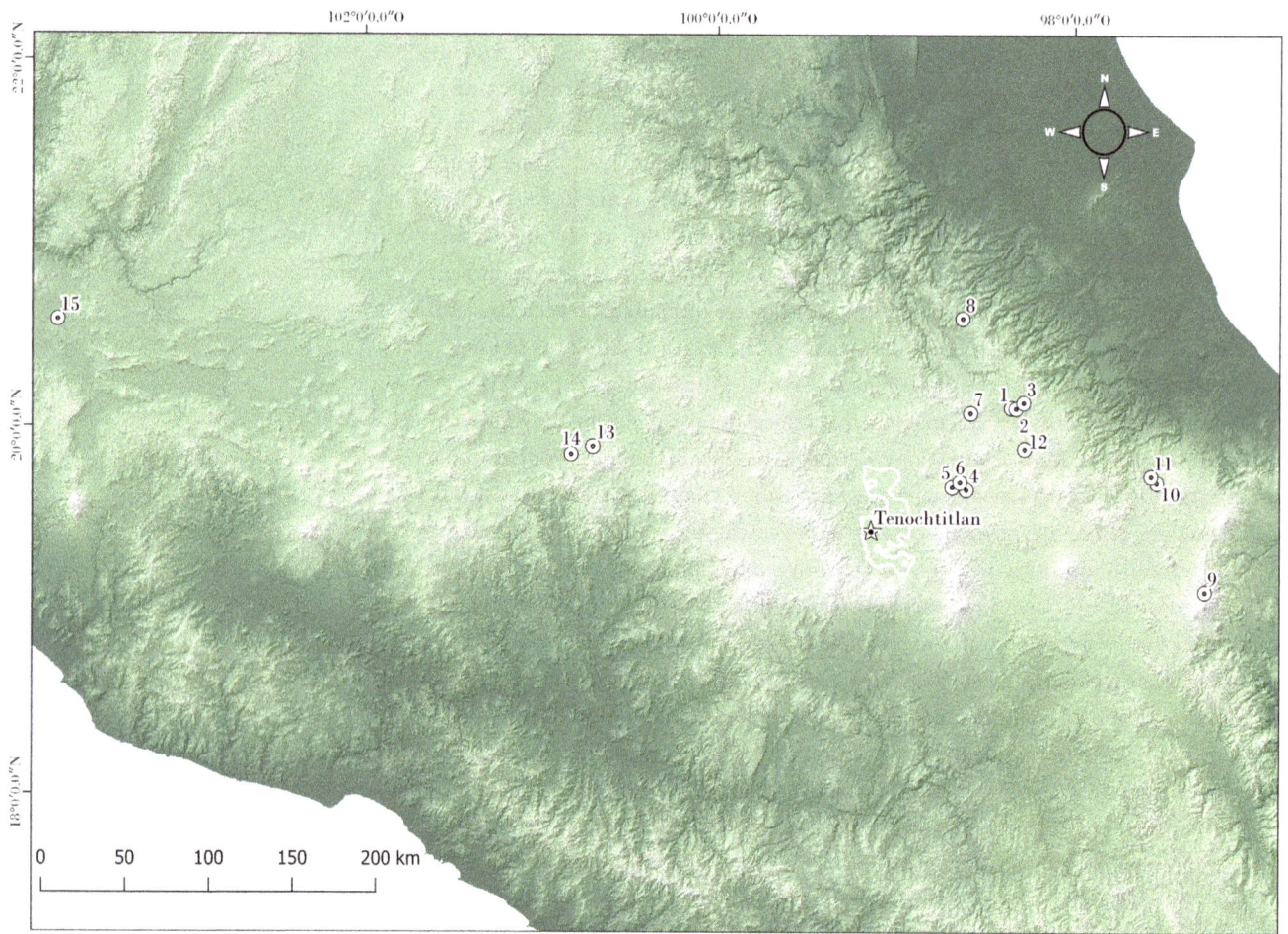

Figure 4. Obsidian sources analyzed for this research: Tulancingo (1), El Abra (2), Tepalzingo (3), Otumba-Malpaís (4), Otumba-Buenavista (5), Otumba-Ixtepec (6), Sierra de Pachuca/Sierra de las Navajas (7), Atopixco (8), Pico de Orizaba-Ixtetal (9), Oyameles (10), Zaragoza (11), Tres Cabezas (12), Ucareo (13), Zinapécuaro (14), Ahuisculco (15). Map by Víctor García Gómez.

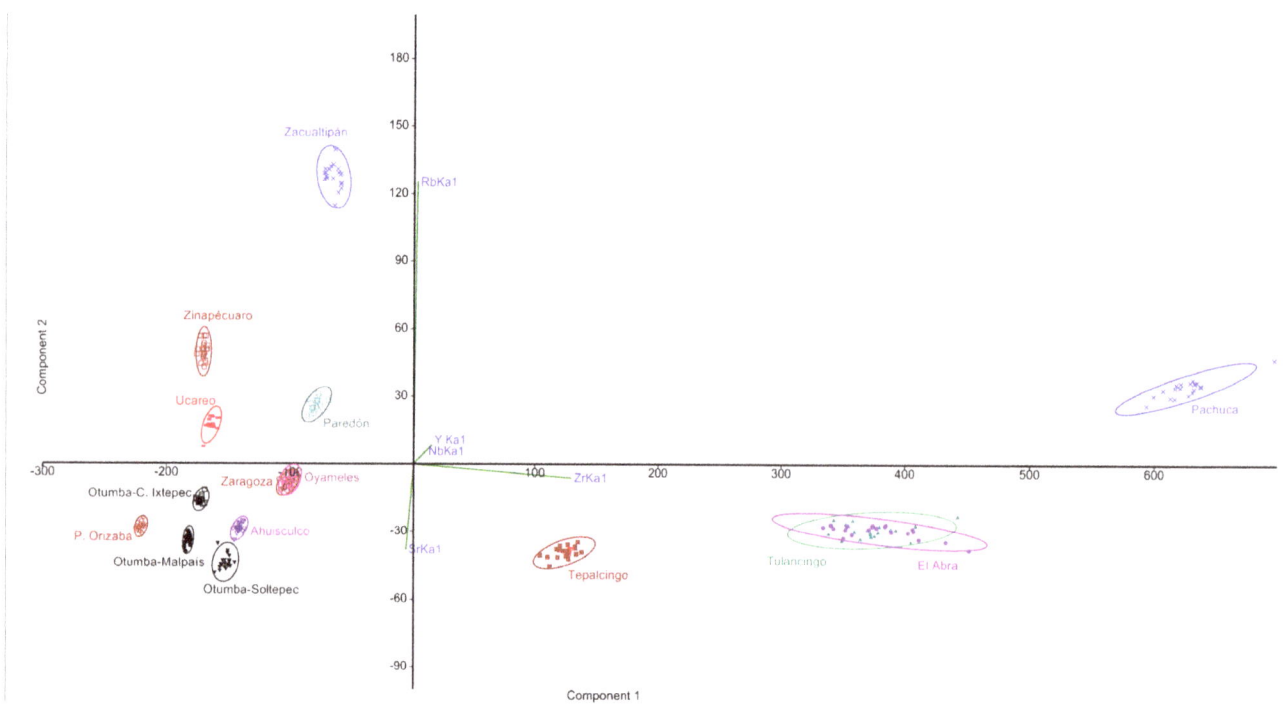

Figure 5. Principal Component Analysis (PCA) of the obsidian sources. Graphic by Guillermo Acosta Ochoa.

Table 2. XRF results of the 10 elements analyzed (values on ppm).

ID Sample	Source	Mn	Fe	Zn	Ga	Th	Rb	Sr	Y	Zr	Nb
10_168842_10	Sierra de Pachuca	938	15361	218	26	19	202	3	109	914	91
10_168842_B	Sierra de Pachuca	1003	15349	216	27	20	206	3	110	907	89
10_265272	No id.	1008	13264	156	8	31	99	4	58	353	44
10-168825_1	Sierra de Pachuca	1026	15414	211	23	18	193	3	107	854	87
10-168825_2	Sierra de Pachuca	1079	17268	258	26	18	192	7	107	870	87
10-168842	Sierra de Pachuca	890	14777	204	22	17	187	3	100	840	85
10-168842_3	Sierra de Pachuca	1062	16034	227	27	18	205	2	116	945	94
10-168842_4	Otumba-Buenavista	342	7752	44	18	9	115	113	22	121	13
10-168842_11	Sierra de Pachuca	849	14516	195	23	20	190	2	107	865	83
10-220244_2	Sierra de Pachuca	1040	14914	245	22	15	192	3	106	845	85
10-220245_1	Otumba-Buenavista	344	8126	48	17	11	119	114	20	135	13
10-220245_3	Otumba-Buenavista	427	8051	51	19	10	120	106	22	126	13
10-220245_3B	Otumba-Buenavista	363	8130	43	18	9	117	112	22	129	12
10-220344	Sierra de Pachuca	1077	14542	208	24	17	193	4	105	831	84
10-220344_1	Ucareo	174	7039	41	19	11	135	13	22	107	12
10-265278_4	Pico de Orizaba	382	5274	40	27	8	135	53	18	83	11
10-265278-1	Pico de Orizaba	327	5389	41	29	9	153	54	19	75	10
10-265278-6	Pico de Orizaba	263	4841	36	23	8	126	48	17	69	10
214.1	Sierra de Pachuca	883	14522	1092	22	18	190	3	107	855	84
3035	Otumba-Buenavista	397	9364	118	20	12	126	125	21	137	13
V567-10783	Sierra de Pachuca	961	14747	199	25	18	177	4	99	796	76

Results

The Principal Component Analysis (PCA) of the obsidian artifacts from the Great Temple, compared with the geological samples, allowed us to determine that 5 of the 10 elements analyzed (Y, Rb, Sr, Zr, and Nb) are the most significant in defining provenance (Figures 6 and 7); while principal components 1 and 2 include 99.5% of the variance (Table 2).

This preliminary analysis allowed us to determine that most of the samples belong to the main deposits of the Basin of Mexico (Otumba and Sierra de Pachuca). However, it was also possible to identify two other sourcing areas with more accuracy by another statistical analysis—Hierarchical Cluster Analysis. This analysis allowed us to appreciate that the duck head pendants (objects 10-265278-1, 10-265278-4, and 10-265278-6) came from Pico de Orizaba in Veracruz, while one helical bead (object 10-220344_1) was made with obsidian from Ucareo in Michoacan. The results are summarized in Table 3.

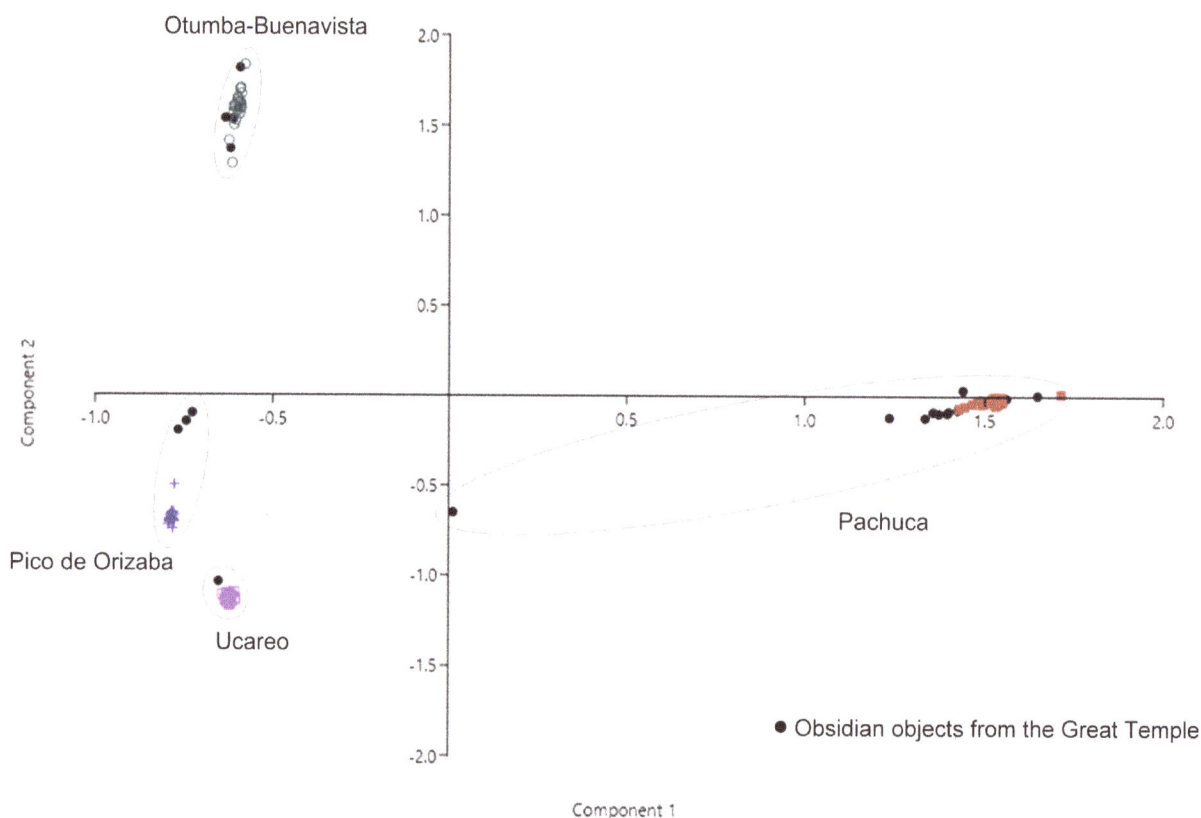

Figure 6. Results of the obsidian sources identified on the objects from Tenochtitlan by PCA. Graphic by Guillermo Acosta Ochoa.

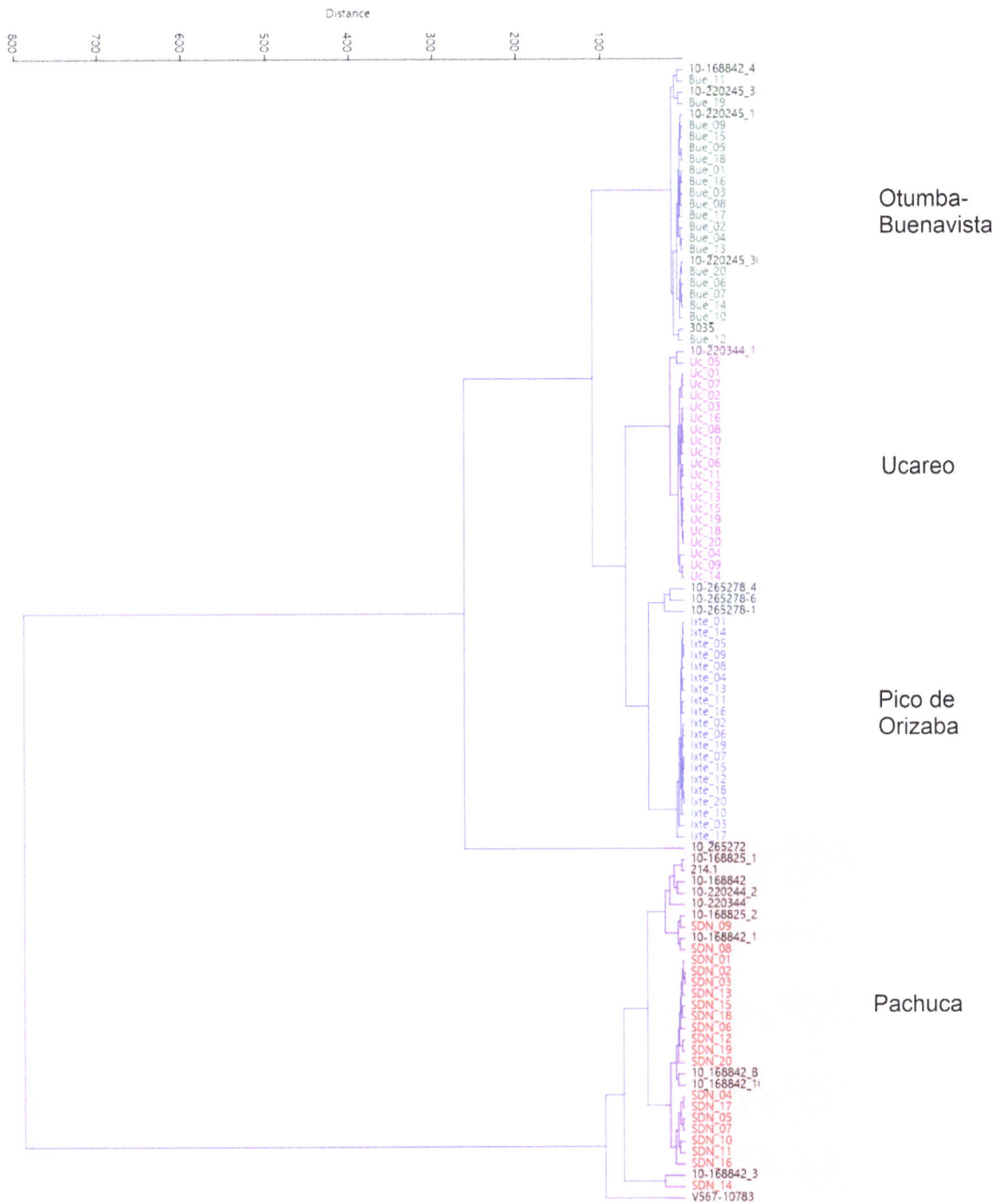

Figure 7. Hierarchical Cluster Analysis of the obsidian sources identified on the objects from Tenochtitlan. Graphic by Guillermo Acosta Ochoa.

Table 3. XRF results on the identified obsidian sources from the Great Temple of Tenochtitlan.

ID Sample	Object	Offering	Structure	Stage	Ruler	Color	Obsidian source
10-168825-1	Funerary urn	34	Great Temple	II	Acamapichtli, Huizilíhuitl, and Chimalpopoca	Green	Sierra de Pachuca
10-168825-2	Urn lid						Sierra de Pachuca
10-168842	Duck head pendant	14		IVb	Axayácatl	Green	Sierra de Pachuca
10-168842-B							Sierra de Pachuca
10-168842-3							Sierra de Pachuca
10-168842-4							Sierra de Pachuca
10-168842-10							Sierra de Pachuca
10-168842-11							Sierra de Pachuca
10-220244-2	Helical bead	39		II	Acamapichtli, Huizilíhuitl, and Chimalpopoca	Gray	Sierra de Pachuca
10-220245-1	Pumpkin bead					Reddish-brown	Otumba Buenavista
10-220245-3							Otumba Buenavista
10-220245-3B							Otumba Buenavista
10-220344	Helical bead					Green	Sierra de Pachuca
10-220344-1	Helical bead						Ucareo
10-265272	Small cylindrical scepter with globular top	V	House of the Eagles	V	Tízoc	Golden	Sierra de Pachuca
10-265278-1	Duck head pendant					Gray	Pico de Orizaba
10-265278-4							Pico de Orizaba
10-265278-6							Pico de Orizaba
214.1	Broken giant cylindrical scepter with globular top					Golden	Sierra de Pachuca
3035	Globular bead	34	Great Temple	II	Acamapichtli, Huizilíhuitl, and Chimalpopoca	Reddish-brown	Otumba Buenavista
V567-10783	Broken giant cylindrical scepter with globular top	V	House of the Eagles	V	Tízoc	Golden	Sierra de Pachuca

Conclusions

Obsidian was a resource of enormous importance to precolumbian societies. The main political centers in the Basin of Mexico, from Teotihuacan to Tenochtitlan, tried to control the exploitation of the local deposits, craft specialization, and distribution of this resource (Carballo 2011; Cobean 2002, Pastrana Cruz 1991, 2007; Kwoka and Shackley 2019). The obsidian had different purposes and values (Furholt, this volume), from the more general aspects of craft and food production to its consumption in aesthetic, ritual, and military realms (Matos Moctezuma 1989; Levine and Carballo 2014). Also, the diversity of goods crafted with obsidian sometimes marked social distinctions within and among producers and consumers (Werra et al., this volume), and their circulation was controlled through state channels such as markets and specialized traders as in other precolumbian groups (Nash, this volume).

In the case of the Triple Alliance, we have noted that obsidian does not appear as a tributary good in the *Matricula de Tributos*, despite the importance of ensuring its access to a militaristic lordship such as Tenochtitlan (Pastrana Cruz 1991: 86). The likely reason for this is that the Triple Alliance had ensured its access directly by controlling the extraction directly in the mines, as seems to have happened at Sierra de Pachuca, which explains the wide dominance of green obsidian in the Late Postclassic sites of the Basin of Mexico. Although this applies to Stages IVb and V of the Great Temple (Offerings V and 14), thus corresponding to the governments of Axayácatl and Tízoc, it is not applicable for Stage II (Offerings 34 and 39), since they correspond to the pre-imperial period when Tenochtitlan was subject to the Tepanec realm of Azcapotzalco.

Some authors consider that green obsidian owes its wide distribution during the Late Postclassic period throughout the Mesoamerican area to a distribution system resulting from the Pochteca or long-distance trade (Clark 1988; Pastrana Cruz 1991), which would explain the presence of green obsidian from Sierra de Pachuca and reddish obsidian from Otumba in the offerings of Stage II at the Great Temple, regardless of whether they were obtained as raw material or finished product, given the high degree of craft specialization required. In the case of the obsidian from Otumba, this source was dominated by the ruler of Texcoco since the thirteenth century; thus, its acquisition during the pre-imperial period at Tenochtitlan could have been through the market system in the Basin of Mexico. It is worth noting, however, that the obsidian used for manufacturing the pumpkin beads (10-220245-1, 10-220245-3, 10-220245-3B) and globular beads (3035) is the reddish-brown from Otumba, an uncommon obsidian color employed to craft lithic tools in central Mexico (Glascock et al. 1994; Cobean 2002). On the other hand, we do not know if these objects were manufactured in Tenochtitlan or gathered as finished artifacts since Cynthia Otis Charlton (1993) identified specialized obsidian lapidary workshops in Otumba that include the reddish obsidian color. This could suggest a prior relationship between Tenochtitlan and Texcoco; perhaps the objects were used as war prizes in the conquest of Otumba during the reign of Huitzilíhuitl or as part of the Tepanec War campaign against Texcoco in AD 1403, a time when the Mexica were vassals of Azcapotzalco.

Among the foreign obsidian materials, we identified two deposits outside the Basin of Mexico (Figure 8): Pico de Orizaba (Offering V) and Ucareo (Offering 39). The Pico de Orizaba obsidian was used to manufacture three duck head pendants in translucent gray obsidian. The unique color of this obsidian led us to consider Paredon or Pico de Orizaba as the most probable sources. XRF analysis confirmed the provenance from the Veracruz deposit. Atypical among the obsidian artifacts from the Basin of Mexico, this material is unique among the Tenochcan assemblage. Interestingly, the objects crafted with this obsidian were deposited inside an offering during the five years of the reign of Tízoc, a very complicated sociopolitical period in Tenochtitlan with scarce military achievements, which triggered the poisoning and subsequent death of this ruler as part of a conspiracy made by Techotlala, lord of Iztapalapa, and Maxtlaton, lord of Tlachco (Torquemada 1969: 184–185). This event must have caused a strong conflict within the Mexica government, altering the sociopolitical alliances as well as the areas of obtaining stone materials for the government of his successor Ahuítzotl. We can surmise this since the chemical compositions and diversity of precious raw materials, such as turquoises and various green stones, changed drastically in comparison with the objects of previous stages (Melgar Tísoc 2014: 198–209; Melgar Tísoc and Solís Ciriaco 2010: 120; Ruvalcaba Sil et al. 2013: 170–172).

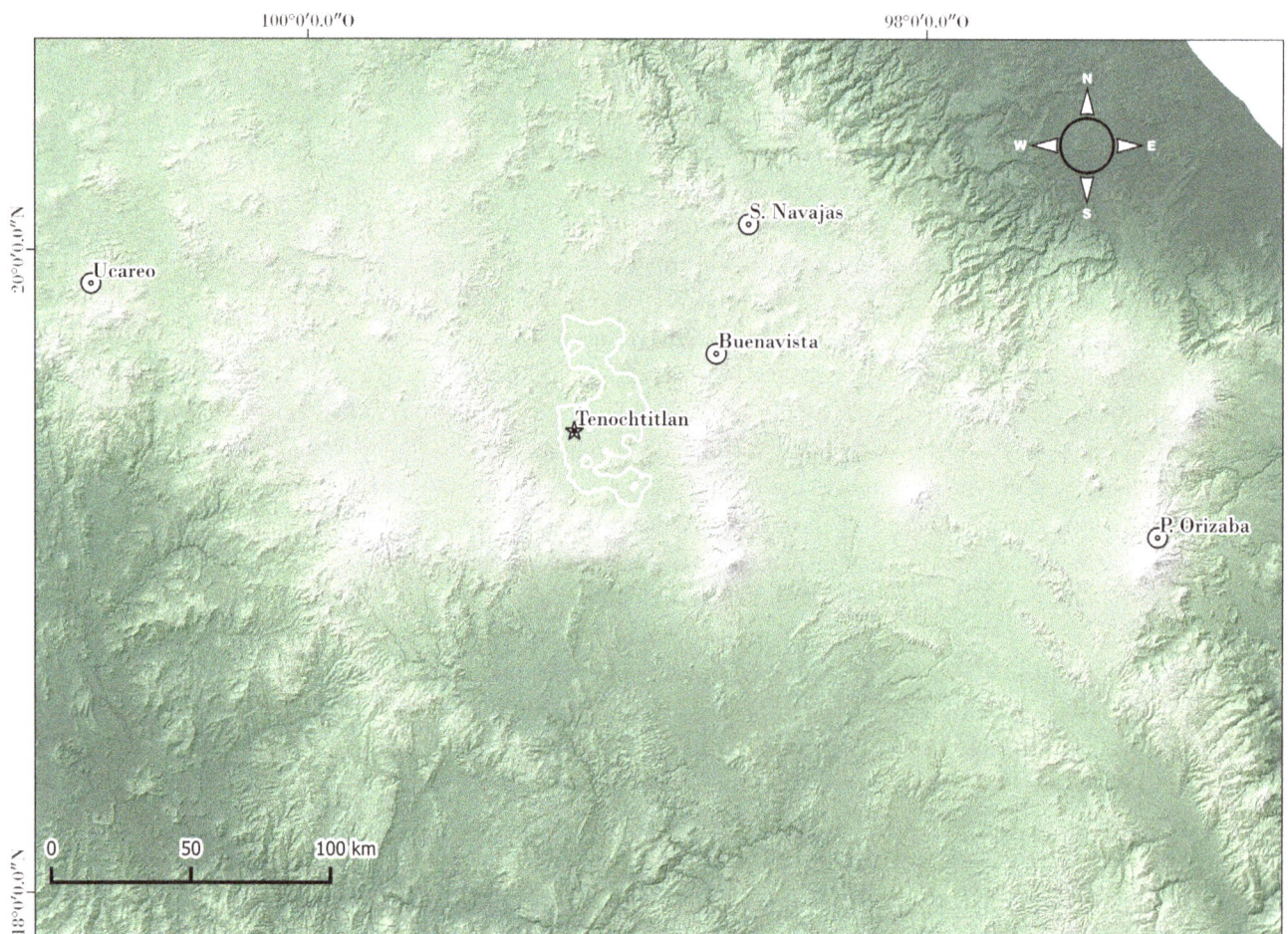

Figure 8. Sources identified on the objects from Tenochtitlan. Map by Víctor García Gómez.

However, the emergence of obsidian from Pico de Orizaba is not entirely unprecedented considering that the Orizaba region was of strategic importance to the Triple Alliance. Studies of obsidian mines in the Ixtetal valley on the eastern slope of Pico de Orizaba show intensive exploitation of prismatic and bifacial blade cores (Stocker and Cobean 1984; Pastrana Cruz 1986, 1991). The interest of the Triple Alliance in this deposit was most likely not in obtaining obsidian for Tenochtitlan but instead for the military garrisons located about 20 km from the mines. While Moctezuma I undertook the military campaigns toward the Gulf Coast, conquering Orizaba (Ahuilizapan) among other towns in the region (Durán 2006: 177–181), Tízoc, following the rebellions of towns conquered by his predecessors, probably sought to maintain the territory more than to expand it—and Pico de Orizaba mines must have been strategic during his reign.

Finally, and as an atypical element within the identified deposits, there is a dark obsidian helical bead (10-220344-1) from Offering 39 during Stage II of the Great Temple (AD 1375–1426).

After the green obsidian from Sierra de Pachuca, the obsidian from Zinapécuaro-Ucareo is considered to be one of the main deposits that supplied central Mexico (Cobean 2002; Healan 1998), particularly during the Formative period (Boksenbaum et al. 1987). After losing the hegemony of Teotihuacán, Ucareo obsidian regains importance again as a pan- regional foreign good, especially during the Late Classic and Epiclassic when Ucareo obsidian becomes the dominant source for sites associated with Coyotlatelco ceramics such as Tula, Cerro Portezuelo, Azcapotzalco, and the Toluca Valley (Healan 1998; Parry and Glascock 2013; Sugiura et al. 2018). For the Postclassic period, obsidians from Sierra de Pachuca and Otumba are again relevant, while materials from Ucareo are unusual probably because of the dominance of the mine by the Tarascan Empire of Tzintzuntzan and the subsequent conflict with the Triple Alliance (Healan 1998; Hernández 2000). This leads us to consider that this artifact is a relic, obtained directly or indirectly by the Tenochcan people either by looting sites in the Basin of Mexico with Coyotlatelco occupation (such as Tula or Azcapotzalco) or as a gift or war prize. The use of relics from previous cultures and sites has been widely documented in the Great Temple (Matos Moctezuma 1988; López Luján 1993; Melgar Tísoc et al. 2018; Monterrosa Desruelles 2018), but this piece made with obsidian from Ucareo is the first confirmed material from that region among the assemblage of the Great Temple of Tenochtitlan.

Finally, this study shows the importance of the multidisciplinary study of the lapidary objects from the Great Temple and opens new questions for future research for other materials and lapidary objects from places and/or groups beyond the imperial borders that supposedly should not be in the offerings of the Great Temple of Tenochtitlan.

Acknowledgments

This research was funded by *Consejo Nacional de Ciencia y Tecnología* (National Council of Humanities, Science, and Technology; CONACYT) through *Ciencia Básica* (Basic Science) grant number CB-283896, and the permit of the Council of Archaeology from the *Instituto Nacional de Antropología e Historia* (National Institute of Anthropology and History;

INAH) to perform the analysis on archaeological objects. We would like to thank Patricia Ledesma Bouchan, director of the Great Temple Museum, and the staff of this museum for the facilities to analyze the obsidian objects from Tenochtitlan. We also thank Paulina Jiménez for helping us with the translation of the text.

References

Acosta Ochoa, Guillermo, Víctor Hugo García Gómez, and Juan Rodrigo
Esparza López
  2015   Análisis de procedencia de obsidianas de la Depresión Central de Chiapas.
  In *XXVIII Simposio de Investigaciones Arqueológicas en Guatemala, 2014*, edited by
  B. Arroyo, L. Méndez Salinas, and L. Paiz, pp. 901–908. Museo Nacional de
  Arqueología y Etnología, Guatemala.

Athié Islas, Ivonne
  2001   La Obsidiana del Templo Mayor de Tenochtitlan. Bachelor's thesis,
  Archaeology Department, National Institute of *Anthropology* and History (*INAH*),
  Mexico City.

Boksenbaum, M. W., P. Tolstoy, G. Harbottle, J. Kimberlin, and M. Neivens
  1987   Obsidian Industries and Cultural Evolution in the Basin of Mexico before
  500 BC. *Journal of Field Archaeology* 14: 65–75.

Braswell, Geoffrey E.
  1997   El intercambio prehispánico en Yucatán, México. In *X Simposio de
  Investigaciones Arqueológicas en Guatemala, 1996*, edited by J. P. Laporte and H.
  Escobedo, pp. 595–606. Museo Nacional de Arqueología y Etnología, Guatemala.

Carballo, David
  2011   Obsidian and the Teotihuacan State. UNAM, University of
  Pittsburgh, Mexico.

Cobean, Robert H.
  2002   *Un mundo de obsidiana: Minería y comercio de un vidrio volcánico en el México
  antiguo*. INAH, Mexico City.

Charlton, Cynthia Otis
  1993   Obsidian as Jewelry: Lapidary Production in Aztec Otumba, Mexico.
  Ancient Mesoamerica 4(2): 231–243.

Clark, John
  1988   *The Lithic Artifacts of La Libertad, Chiapas, Mexico: An Economic Perspective*.
  New World Archaeological Foundation, Brigham Young University, Provo, Utah.

Durán, fray Diego
  2006   *Historia de las Indias de Nueva España e Islas de la Tierra firme*.
  Porrúa, Mexico.

García Gómez, Víctor Hugo
2018   Procedencia e intercambio de obsidiana en la Cuenca de México en el Holoceno Medio (6000–4000 a.n.e.): el caso de San Gregorio Atlapulco, Xochimilco. Master's thesis, Anthropology Research Institute, UNAM, Mexico City.

Glascock, Michael D.
2011   Comparison and contrast between XRF and NAA: used for characterization of obsidian sources in central Mexico. In *X-Ray Fluorescence Spectrometry (XRF) in Geoarchaeology*, edited by M. Steven Shackley, pp. 161–192. Springer, New York.

Glascock, Michael, and Hector Neff
1999   Sources of obsidian artifacts from Templo Mayor. Report on file at the Missouri University Research Reactor (MURR). Columbia, Missouri.

Glascock, Michael, Hector Neff, Joaquín García Bárcena, and Alejandro Pastrana
1994   La obsidiana 'Meca' del centro de México, análisis químico y petrográfico. *TRACE* 25: 66–73.

Healan, Dan
1998   La cerámica Coyotlatelco y la explotación del yacimiento de obsidiana de Ucareo-Zinapécuaro. In *Génesis, culturas y espacios en Michoacán*, edited by Veronique Darras and Charlotte Arnauld, pp. 101–111. Centro de estudios mexicanos y centroamericanos, Mexico City.

Hernández, Christine
2000   A History of Prehispanic Ceramics, Interaction, and Frontier Development in the Ucareo-Zinapécuaro Obsidian Source Area, Michoacan, México. PhD dissertation, Anthropology Department, Tulane University, New Orleans.

Jack, R. N., and R. F. Heizer
1968   Finger-Printing of Some Mesoamerican Obsidian Artifacts. *Contributions of the University of California Archaeological Research Facility* 5: 81–100.

Kwoka, Joshua J., and M. Steven Shackley
2019   Technological Analysis and Source Provenance of Obsidian Artifacts from a Sun Pyramid Substructure Cache, Teotihuacán, Mexico. *Latin American Antiquity* 30: 205–210.

Levine, Marc, and David Carballo (editors)
2014   *Obsidian Reflections: Symbolic Dimensions of Obsidian in Mesoamerica*. University Press of Colorado, Boulder.

López Luján, Leonardo

    1993   *Las ofrendas del Templo Mayor de Tenochtitlan.* INAH, Mexico City.

Matos Moctezuma, Eduardo

    1988   *The Great Temple of the Aztecs. Treasures of Tenochtitlan.* Thames and Hudson, London.

    1989   *Guía oficial. Templo Mayor.* INAH-Salvat, Mexico City.

Melgar Tísoc, Emiliano Ricardo

    2014   Comercio, tributo y producción de las turquesas del Templo Mayor de Tenochtitlan. Master's thesis, Anthropological Research Institute, UNAM, Mexico City.

Melgar Tísoc, Emiliano Ricardo, and Reyna Beatriz Solís Ciriaco

    2009   Caracterización de huellas de manufactura en objetos lapidarios de obsidiana del Templo Mayor de Tenochtitlan. *Arqueología* 42: 118–134.

    2010   Manufacturing Techniques of the Turquoise Mosaics from the Great Temple of Tenochtitlan, Mexico. In *LASMAC: 2nd Latin American Symposium on Physical and Chemical Methods in Archaeology, Art and Cultural Heritage Conservation. Symposium on Archaeological and Art Issues in Materials Science,* edited by José Luis Ruvalcaba Sil, Javier Reyes Trujeque, Jesús Arenas Alatorre, and Adrián Velázquez Castro, pp. 119–124. Sociedad Mexicana de Materiales-INAH-UNAM-UAC, Mexico City.

Melgar Tísoc, Emiliano Ricardo, Reyna Beatriz Solís Ciriaco, and Hervé Víctor Monterrosa Desruelles

    2018   *Piedras de fuego y agua. Turquesas y jades entre los nahuas.* INAH, Mexico City.

Monterrosa Desruelles, Hervé Víctor

    2018   La manufactura de los objetos de jadeíta verde imperial de las ofrendas del Templo Mayor de Tenochtitlan. *Revista Española de Antropología Americana* 18: 211–231.

Nazaroff, A. J., K. M. Prufer, and B. L. Drake

    2010   Assessing the Applicability of Portable X-ray Fluorescence Spectrometry for Obsidian Provenance Research in the Maya Lowlands. *Journal of Archaeological Science* 37: 885–895.

Obregón Rodríguez, María Concepción

    1995   La zona del Altiplano central en el Posclásico: la etapa de la Triple Alianza. In *Historia Antigua de México. Vol. III: El horizonte Posclásico y algunos aspectos intelectuales de las culturas mesoamericanas,* edited by Linda Manzanilla and Leonardo López Luján, pp. 265–306. INAH-UNAM-Porrúa, Mexico.

Olmo Frese, Laura del
1999   *Análisis de la ofrenda 98 del Templo Mayor de Tenochtitlan.* INAH, Mexico City.

Parry, William, and Michael Glascock
2013   Obsidian Blades from Cerro Portezuelo: Sourcing Artifacts from a Long-Duration Site. *Ancient Mesoamerica* 24: 177–184.

Pastrana, Alejandro
1986   El proceso de trabajo de la obsidiana de las minas de Pico de Orizaba. *Boletín de Antropología Americana* 13: 132–145.
1991   Iztepec, Izteyoca e Iztla… distribución mexica de obsidiana. *Arqueología* 6: 85–100.
2007   *La distribución de la obsidiana de la Triple Alianza en la Cuenca de México.* INAH, Mexico City.

Reuther, Joshua D., Natalia S. Slobodina, Jeff Rasic, John P. Cook, and Robert J. Speakman
2011   Gaining Momentum. The Status of Obsidian Source Studies in Alaska: Implications for Understanding Regional Prehistory. In *From the Yenisei to the Yukon Interpreting Lithic Assemblage Variability in Late Pleistocene/Early Holocene Beringia*, edited by Ted Goebel and Ian Buvit, pp. 270–286. Texas A&M University Press, Austin.

Ruvalcaba Sil, José Luis, Emiliano Ricardo Melgar Tísoc, Kilian Laclavetine, Jessica Curado, and Thomas Calligaro
2013   Caracterización y procedencia de piedras verdes de las ofrendas del Templo Mayor de Tenochtitlan. In *Técnicas analíticas aplicadas a la caracterización y producción de materiales arqueológicos en el área Maya*, edited by Adrián Velázquez and Lynneth Lowe, pp. 163–168. UNAM, Mexico City.

Shackley, Steven M.
2011   An Introduction to X-Ray Fluorescence (XRF) Analysis in Archaeology. In *X-Ray Fluorescence Spectrometry (XRF) in Geoarchaeology*, edited by M. S. Shackley, pp. 7–44. Springer, New York.

Solís Ciriaco, Reyna Beatriz
2018   Esferas de producción de los objetos de piedra verde procedentes de las estructuras aledañas al Templo Mayor de Tenochtitlan. *Revista Española de Antropología Americana* 18: 233–249.

Speakman, Robert J.
2012   *Evaluation of Bruker's Tracer Family Factory Obsidian Calibration for Handheld Portable XRF Studies of Obsidian.* Bruker AXS, Kennewick, Washington.

Speakman, Robert J., and M. Steven Shackley

2013   Silo Science and Portable XRF in Archaeology: A Response to Frahm. *Journal of Archaeological Science* 40: 1435–1443.

Stocker, Terrace, and Robert H. Cobean

1984   Preliminary Report on the Obsidian Mines at Pico de Orizaba, Veracruz. In *Prehistoric Quarries and Lithic Production*, edited by J. Ericson and B. Purdy, pp. 83–95. Cambridge University Press, Cambridge.

Sugiura, Yoko, Gustavo Jaimes, Shigeru Kabata, and Michael D. Glascock

2018   La obsidiana como un bien de intercambio entre el valle de Toluca y sus regiones circunvecinas durante el Clásico. *Anales de antropología* 52(2): 55.

Torquemada, Juan de

1969   *Monarquía Indiana*. Porrúa, Mexico.

Velázquez Castro, Adrián, and Emiliano Ricardo Melgar Tísoc

2014   Producciones palaciegas tenochcas en objetos de concha y lapidaria. *Ancient Mesoamerica* 25(1): 295–308.

Vilanova de Allende, Rodrigo

2002   Asentamientos de la Triple Alianza en su frontera norte: el Valle del Mezquital. *Arqueología* 28: 93–104.

CHAPTER 7

Getting to the Point: Wari Obsidian Distribution, Reduction, and Use on the Southern Frontier, Moquegua, Peru

DONNA J. NASH

Donna J. Nash Arizona State University, Arizona, USA (donna.j.nash@asu.edu)

Abstract

Obsidian was mobilized in a special way by the Wari Empire (ca. 600–1100 CE) during the Andean Middle Horizon. Previous research has identified the sources of obsidian that were used and distributed through imperial channels of circulation. Geochemical techniques can and have tracked the movement of obsidian across the Andes, but they cannot elucidate how this resource was managed and processed as it moved between quarries and its eventual users. Materials recovered from Cerro Baúl and Cerro Mejía, two settlements in Moquegua on the southern frontier, provide a perspective from the final link in this chain and demonstrate that some provinces in the empire received preforms, which were distributed to all provincial "citizens." The Moquegua province provides a unique opportunity to study the imperial distribution system because very little obsidian entered the region prior to Wari incursion, when the distinctive and diagnostic "Classic Wari Laurel Leaf Point" appeared in Moquegua and in many regions throughout the Peruvian Andes. In this chapter, I suggest that this diagnostic form derives its general shape from the processing of cores into preforms, which were distributed throughout the empire, whereas smaller triangular points were made from flakes removed from preforms used as multidirectional cores through efficient bifacial reduction. I use evidence from two production locales and household assemblages to demonstrate the manner in which preforms became points, how obsidian was distributed to colonists, and how it was used by those living on the southern frontier of the Wari Empire.

Introduction

At the time of European contact, the Andean region of South America, dominated by the Inka Empire (ca. 1350–1532 CE), was largely a stone age society. People living in earlier eras likewise relied heavily on chipped stone for agriculture, cutting, chopping, crafting, and warfare. Since rock and minerals of different sorts were abundant and readily available, domestic assemblages consist primarily of expedient rather than curated tools. Assemblages in the earliest empire of the Andes, the Wari (ca. 600–1100 CE), were no exception. Across the 1100 km breadth of the polity's domain from Cajamarca in the north to Moquegua in the south, the massive landscape transformations they directed, which sculpted thousands of hectares of rugged mountainside into irrigated, terraced fields, were achieved with stone hoes. The urban, cosmopolitan capital exceeded 15 km.[2] The ruins of the vast city are in Ayacucho, Peru, and include magnificent megalithic finished stone blocks, multi-story masonry walls, and an intricate subterranean system of channels and drains. Provincial centers in several regions, such as Cusco and Huamachuco, share these features and were built with stone tools of no distinct forms.

The exception to all this expediency in the vast Wari Empire are a distinctive and diagnostic type of obsidian point, which was not produced by their contemporaries: the Moche and Tiwanaku. It was not a resource present in all regions. The special characteristics of obsidian were valued, and people transported it over long distances. Obsidian was mobilized in a special way by the Wari Empire (Burger et al. 2000). Previous research has identified the sources of obsidian (Tripcevich and Contreras 2013) that were used and distributed through imperial channels of circulation and that may have extended beyond the empire's boundaries and supplied exchange partners beyond its frontiers, such as Tiwanaku (Williams et al. 2012).

Geochemical techniques can track the movement of obsidian across the Andes, but it cannot elucidate how this resource was managed and processed as it traveled between locales of acquisition and its eventual users. Materials recovered from Cerro Baúl and Cerro Mejía—two settlements in Moquegua on the southern frontier of the empire, more than 500 km from the capital—form a sizable assemblage to make inference about the distal links in this chain. Items from households, ritual offerings, and two production locales demonstrate that the frontier province of Moquegua received preforms, rather than nodules with cortex, for use by imperial "citizens" (people who owed tribute of some form directly to the state, and perhaps some of those who did so indirectly through one of its clients). Most items that would be considered "Classic Wari Laurel Leaf Points" are bifaces that derive their general shape from the processing of cores into preforms, whereas smaller triangular points are typically made from flakes that could have been removed from preforms through efficient bifacial reduction. This manner of distribution may have been typical of multiple provinces, especially those without obsidian sources or a history of obsidian use before Wari incursion.

Obsidian was probably processed in the Wari capital, and perhaps a few other facilities near sources, before it was distributed to provincial elites. At this time, information about processing near quarries is not available. In any case, it may have been regulated in some manner because one type of Wari obsidian point has a diagnostic shape found throughout the empire. I suggest that these Classic Wari Laurel Leaf Points are diagnostic bifaces because they derive their general form from the processing of cores or large flakes into preforms for distribution. On the other hand, smaller triangular points, also fashioned from obsidian, were made from flakes removed from preforms through efficient bifacial reduction. These are more variable in form and resemble local points made from other materials. In this chapter, I describe the process of obsidian reduction in Wari's southern province. I detail the evidence from the palace on Cerro Baúl, as it may be a Wari technology. Studies of the Wari political economy have focused on decorated pottery and its use for feasting. Obsidian, acquired and distributed through formal imperial channels, probably reached a greater proportion of the populace than Wari-decorated pottery, at least in Moquegua. Thus, the study of obsidian and its reduction, distribution, and use offers a new dataset for understanding the articulation between communities and Wari provincial agents throughout the Andes.

Obsidian and the Wari Empire

The Wari polity stretched from Cajamarca in the north to Moquegua in the south, a distance of some 1100 km. The capital was in the central Andes of Ayacucho, Peru. Wari political expansion was materialized during the seventh century CE by the construction of provincial centers, changes in settlements patterns, and the introduction of new artifact styles. Since the Wari Empire is a prehistoric polity, determining its size and strength relies on material culture. In broad strokes, the assemblage associated with the Wari archaeological culture consists of distinctive or canonical architecture, which may be executed using different materials but is consistent in form and the organization of space (Isbell 1991; Nash and Williams 2005, 2009). Ceramic styles are also important but can be equivocal when iconography is the basis for inference and when attributes of technology, locales of production, and manner of distribution are not considered (Nash 2019). Somewhat unusual and perhaps unique to Wari as an Andean complex society is the very diagnostic and widely dispersed Classic Wari Laurel Leaf Point.

This unique aspect of the Wari assemblage begs the question, why? Why are the points diagnostic over such a large area? The relatively similar appearance might be understandable if these items were prestige goods, symbols of power with a limited distribution. Like Badarian maceheads, perhaps they were made as weapons, with fancy versions used to display wealth and authority (Bard 1994). This possibility, however, is not supported by the corpus of representational art. Powerful people or supernatural entities are not depicted holding spears; instead, axes and spear-throwers are common. For example, one metal figure holds an axe in the right hand and a shield in the left (Bergh 2012: 30, Figure 19). Decorated pottery from Conchopata depicts a figure holding an atlatl in the right hand and a diagonally wrapped staff in the left (Bergh 2012: 102, Figure 75c). An impressive metal figurine from Pikillacta shows a human figure seemingly launching an atlatl dart through the air (Bergh 2012: 237, Figure 226A). People are depicted with smaller dark points at the ends of arrows and atlatl darts both on textiles (Bergh 2012: 178,

Figure 1. Map of the Wari Empire with sources of obsidian and the location of Cerro Baúl. Location of obsidian sources based on Tripcevich and Contreras 2013.

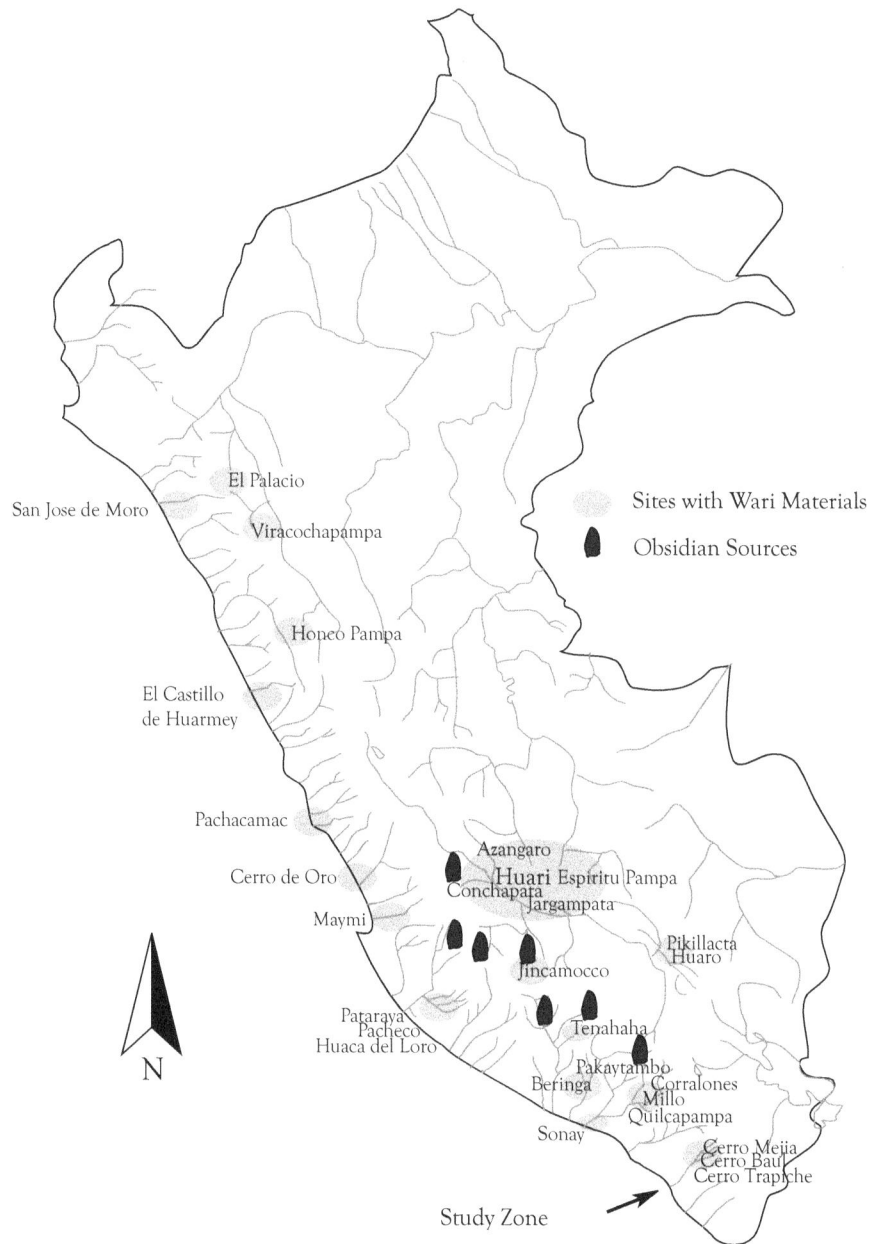

Figure 168) and pottery (Bergh 2012:130, Figure 103). Smaller triangular points are more suitable for darts or arrows.

Wari Laurel Leaf Points are typically too large for darts or arrows (Figure 2, top row). Shott (1997: 94) found that the example of a dart from Nasca (a region within the Wari domain), included in his comparison of darts and arrows, was a small outlier, probably due to its use of a reed main shaft and its diminutive overall length of 400 mm, which exemplifies continuity with the design of arrows. Obsidian points of all forms, with and without evidence of use, are common inclusions in ritual deposits (Figure 2, middle row; see Nash and deFrance 2019). Outside these contexts, retouch flakes affiliated with edge maintenance are found in domestic settings and clustered in hearth ash (e.g., Figure 2K), which indicates their roles in quotidian household activities. Many points exhibit asymmetrical usewear consistent with cutting or shaving (Figure 1.2I), although some examples could have been used for darts (Figure 1.2J). In any case, obsidian was not reserved for weaponry. If that were the case, then one should conclude that Cerro Mejía was a community of soldiers, each of whom maintained their own weapons. I prefer a scenario where women used obsidian to perform domestic tasks and sharpened their tools as needed.

Despite Burger's early revelations about the relationship between obsidian and the Wari polity, most scholars focus their attention on ceramic styles to make interpretations. Geochemical sourcing has shown that both pottery of high quality—which closely

Figure 2. Types of obsidian artifacts found at Wari-affiliated sites in Moquegua. A–D are examples of Classic Wari Laurel Leaf Points. The common forms found in houses of the Baúl-Mejía complex overlap with those found at Tiwanaku sites. Type 1 (H) is rarely made from obsidian and is associated with Tiwanaku (e.g., Klink and Aldenderfer 2005: Figure 3.5). Most examples are denticulated; the tang is long, wide, often rounded, and the wings extend straight to either side (see also Figure 5J). Type 2 (A–D) is the Classic Wari Laurel Leaf. The base of these can be straight or slightly concave. Type 3 (G) has a tang also, but it is short, has a straight terminus, and the wings angle downward on either side. The points themselves are usually shorter and wider than Type 1, have a different length-width ratio, and are usually made using local white chalcedony. Most examples are not denticulated. Type 4 (E) is the most common obsidian form. It is thin with a concave base and at times it is a minimally shaped flake. A few examples can resemble earlier Archaic-Formative Era Points (e.g., J; see Klink and Aldenderfer 2005: Figure 3.6), but most are different in thickness and cross section. Type 5 (F) is only found on Cerro Baúl and, like Type 1, may actually be a projectile. Most pieces have visible indications of hafting. Reworking and resharpening create a great deal of variation. Standardization is not evident among the current sample.

matches what is found in the imperial core—as well as of more modest quality—which had previously been considered derived imitations—were produced locally using similar materials and techniques (Druc et al. 2020; Williams et al. 2019).

The diagnostic form of some Wari obsidian points could be explained in the same manner; however, sources of obsidian are fewer than sources of clay and provided the empire the opportunity to channel distribution through nodes in the political hierarchy. This type of control can never be absolute (Given 2004). Large, complex polities cannot control the activities of all its citizens; no doubt, some obsidian would have moved outside the imperial network, in ancient black markets supplied by outlaws and smugglers. This caveat as well as the features of Wari Laurel Leaf Points, other obsidian points and tools, and the size and type of debitage all provide clues to understand how obsidian moved through the empire and was processed along the way. Since the Wari Empire could control the movement and processing from the source to its ultimate destination in households to a great degree, it is a significant line of evidence to chart the polity's scope and the strength of the relationship between nodes in the imperial network.

Geochemical analyses have identified the sources of obsidian and variation within them. Of primary importance were Quispisisa (Burger and Glascock 2000; Tripcevich 2007) and Alca (Rademaker 2006), as well as Chivay (Burger et al. 1998; Tripcevich 2007), which was also used by Tiwanaku. In their study at Quispisisa, Tripcevich and Contreras (2013) reported primary and assay flakes on the surface near quarry pits, with little evidence of advanced stages of reduction. Further, Jennings and Glascock (2002) reported little evidence of reduction at sources of Alca obsidian. This suggests that, in most time periods, nodules were removed from sources without extensive shaping or processing; however, it is uncertain how obsidian may have been shaped at nearby settlements in the region during different periods. Settlements covered with obsidian reduction debris have been located in the vicinity but have not yet been systematically studied. This leaves a gap in the circulation process, but inferences can be made based on materials found at different sites.

Studies of obsidian artifacts and production debris from the Wari heartland show that some cores were transported there. For example, cortex was present on flakes and bifaces at Vegachayuq Moqo (Kaplan 2018), a central sector of the capital, and at Conchopata (Bencic 2015), a site 12 km away from the city. The majority of pieces analyzed were more than a quarter inch in length or width, a result of the screen size used during excavations. At Conchopata, Bencic found that 18% of obsidian flakes and shatter had some cortex present ($n = 178$ of 985). At Vegachayuq Moqo, Kaplan found a slightly higher percentage of flakes with cortex among her sample (22.7%; $n = 81$ of 358). This pattern is not replicated in Moquegua, where I only found 5 examples with cortex in the analysis of 2,400 pieces (Cerro Baúl [$n = 333$] and Cerro Mejía [$n = 2,067$], 0.21%; $n = 5$ of 2400). These pieces include preforms, complete points, points in production, reduction waste, and retouch flakes from maintenance (see Table 1), which permit me to propose models of reduction that may explain how obsidian was distributed and used by different groups in the Wari colony of Moquegua.

Table 1. The prevalence of cortex among Wari obsidian assemblages.

Location	Sample size	With Cortex	Percentage	Citation
Vegachayoq Moqo	358	81	22.7%	Kaplan 2018
Conchopata	985	178	18%	Bencic 2015
Cerros Baúl & Mejía	2400	5	0.2%	Nash (this paper)

Obsidian reduction in Moquegua

Located in the Torata tributary of the Osmore drainage, Cerro Baúl and Cerro Mejía are adjacent hills occupied by Wari-affiliated households that form two sectors of a low-density urban center designed to represent dual organization (Figure 3; see Nash 2024 for a more detailed description of this configuration), a trait common to later Andean sites such as Inka Cusco (Bauer 1998). Cerro Baúl is a mesa with monumental architecture on the summit and clusters of modest terrace dwellings on its flanks. Cerro Mejía is a steep, dome-shaped hill that follows the same general pattern. The low swale between the two hills was lush with irrigated agricultural fields. The site was occupied by Wari-sponsored frontier colonists of diverse origins and local people drawn from the middle and coastal valley. The cultural diversity is represented by variations in cooking pots, consumption wares, stone tools, and domestic ritual practices (Nash 2017).

Obsidian has been recovered from every house excavated in the Baúl-Mejía settlement cluster, but the types of items recovered differ between the two sites, and there is some indication that the personnel who reduced obsidian differed. On Cerro Baúl, the sample I discuss comes from the palace, located on the summit's eastern end. Thus far, I have only conducted detailed analysis of a fraction of the site's points. Materials in my sample were recovered from a monumental compound in Sector A, an elite residence or palace (Nash 2015, 2018, 2019; Nash and deFrance 2019; Nash and Williams 2021). The palace had personnel adept at reducing preforms to points, and debitage is present in several rooms. On Cerro Mejía, I excavated the house of a specialist or specialists, who shaped points, possibly in service to their neighbors. This house, Unit 19, is on the southern slope of Cerro Mejía, a terrace dwelling where reduction debris was concentrated in a single small room. Debitage from sharpening or edge maintenance has been found in all houses. The sample from Cerro Mejía includes all of the Wari-affiliated houses excavated between 1999 and 2009.

Points and preforms

Some Classic Wari Laurel Leaf Points, which archaeologists recognize and associate with the Wari polity, are preforms. Preforms can be recognized by irregular profiles along their length and uneven surfaces (Figure 4A), whereas a point or finished tool should

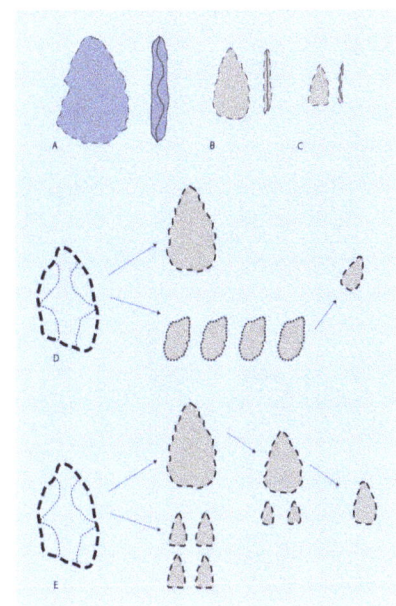

Figure 4. Preforms, points, and reduction practices. The schematic illustrations indicate the features of preforms (A), points reduced from preforms (B), and points created from flakes removed from preforms or large objective pieces (C). The reduction sequence of a preform could produce multiple flakes for use as tools or shaping into small point forms (D). The distribution of preforms through state channels may have involved "skimming" of small flakes along the way so that households received nearly exhausted preforms, small points, or flakes (E).

Figure 3. Map of the Cerro Baúl-Cerro Mejía settlement cluster. Arrows indicate the location of the Sector A Palace on Cerro Baúl and Unit 19 on Cerro Mejía, both of which have evidence for obsidian reduction.

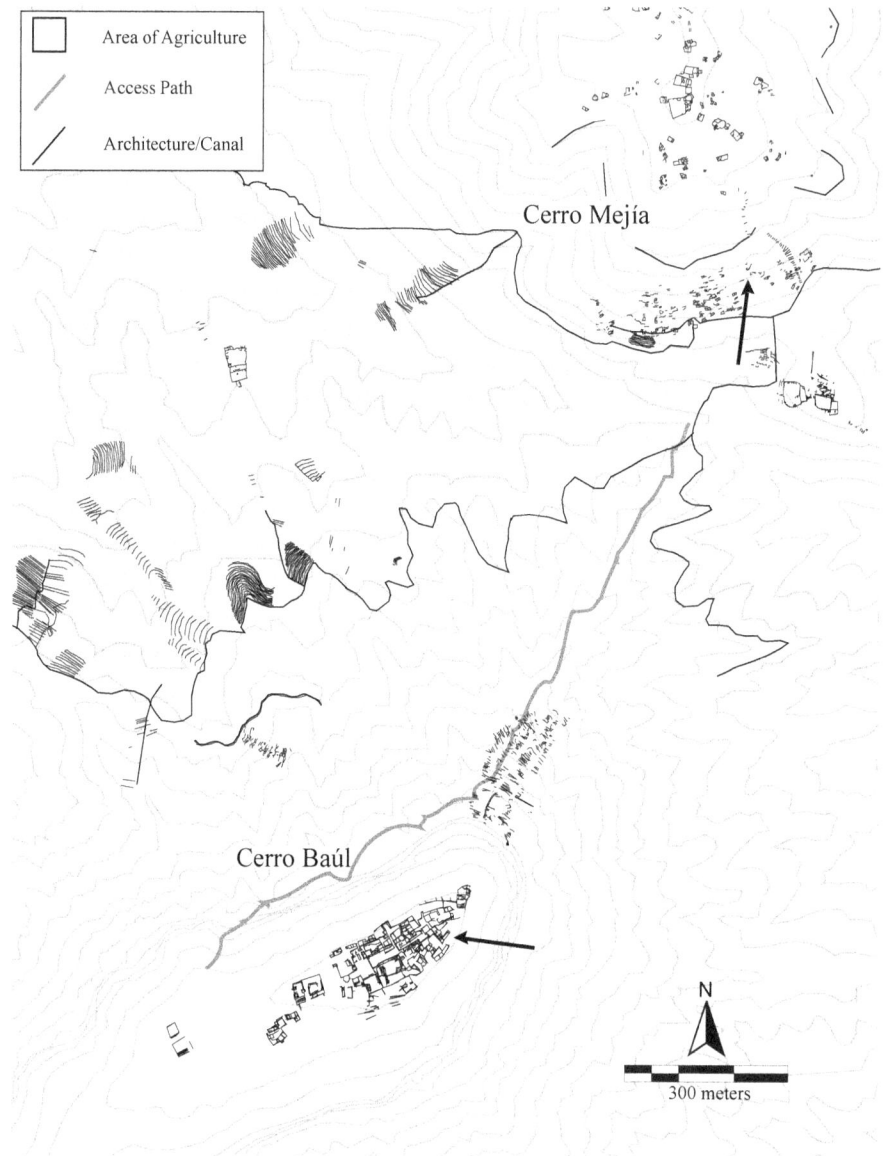

have a relatively straight profile along its length (Figure 4B). Preforms are transformed into points through removing flakes of various sizes. Points reduced from preforms are relatively thin, have straight profiles and maintain the leaf shape (Figure 5C). On the other hand, smaller triangular points, for the most part, can be made from flakes removed from preforms (Figure 4D). They are biproducts derived from preform reduction. Hence, they are slightly bent in profile (Figure 4C). The preforms serve as cores for flake removal and are used as knives when they become too small. This manner of reduction sequence requires skill and technical knowledge. People who do not know how to reduce a preform may use it as a tool. Also, personnel with access to an abundance of obsidian may not choose to reduce it in an efficient manner. Thus, the features of obsidian tools and debitage at a site may provide clues to infer access and participation in the imperial network of distribution.

Figure 5. Examples of obsidian objects found at Cerro Baúl and Cerro Mejía. Two large preforms were found in a construction offering (A–B). Among the assemblage very few pieces with cortex were found (D–E). Laurel Leaf points were not always expertly reduced before they were used but usually retain the general shape of the preforms (C, H, and I). Other point forms are usually less than 30 mm long.

On Cerro Baúl, two large preforms were found, one on top of the other, in a construction offering under the floor of a room in the palace (Figure 5A and B). These pieces are on display at Museo Contisuyo. I was permitted to briefly examine these specimens in the dark exhibition hall. They measure approximately 105 x 66 mm and 95 x 63 mm. The smaller example has large, visible flake scars from the removal of pieces large enough to make small points or other tools. The largest of which measured 30 x 30 mm.

Reduction of preforms to points

The Cerro Baúl palace assemblage provides many examples from which to infer the reduction sequence of preforms, which could be used as multi-directional cores. Flakes removed from preforms were then shaped into triangular points, most of which could be fashioned from a flake 30 by 30 mm in size or less (Figure 5F, G, J, and K). Flakes removed from preforms could also be used for other tasks. I have only observed 5 pieces with cortex from the Baúl-Mejía sample (n = 2400) and suggest that more preforms came to Moquegua without it (Figures 5D and E). It is possible that cortex was removed before preforms entered the Sector A Palace on Cerro Baúl or were distributed to householders on Cerro Mejía. A cache of 93 points was found in 1993 during mapping of Baúl's summit and may have been a location of primary reduction; however, a quarter inch screen was used when this space was excavated in 1997, and these materials have not been analyzed.

Palace artisans used flakes of several shapes to make points. Short, wide flakes were often selected. The bulb of percussion would be oriented to one lateral side of the base (Figure 6). The flake was often polished to create a rough surface resembling frosted glass. Some pieces also exhibit striations. Early in the process, the bulb may have been removed if it was too thick. This appears to have broken some "points in process," as one side is often missing from unfinished points. Detailing of edges was not finished until the bulb was reduced and the base was shaped. Pressure flaking was used to shape flakes into triangular points and maintain obsidian tools. Deer antler tines were found in the palace and were recovered from one household on Cerro Mejía. These items are poorly preserved with severely eroded surfaces and do not permit any definitive conclusion regarding their use as tools.

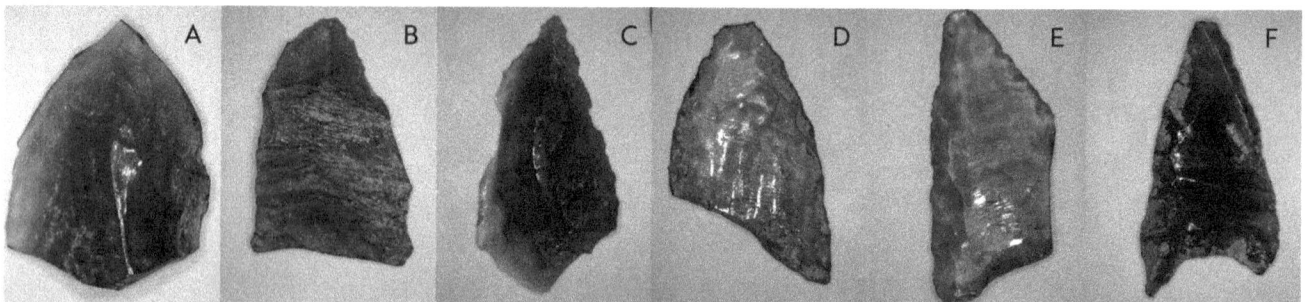

Figure 6. Obsidian points in production. These pictures were taken with a digital microscope; all artifacts shown are less than 30 mm in length. A–C are examples of flakes in early stages of shaping that have been polished to prevent slippage. D–F seem to have been broken while shaping the base, which took place before the two lateral edges were completely finished.

A model of state distribution

The patterns of production and composition of the assemblages in different settings in the Wari colony of Moquegua supports a model of distribution where obsidian originated from state channels of circulation (Table 2). Cerro Mejía has a greater number of pieces; however, the majority are retouch flakes, small thinning or shaping flakes. Points run small. In contrast, specimens from Cerro Baúl include preforms, large and small points, flakes of sufficient size to make points, as well as thinning and shaping flakes. To date, no preforms have been found on Cerro Mejía. In general, more modest houses on the slopes of Cerro Mejía have smaller points—or the only evidence of use in the structure was the presence of retouch flakes created during maintenance. This pattern supports a model of distribution where most obsidian originated from state channels of circulation.

I suggest that large preforms with little to no cortex were delivered to elite governors on Cerro Baúl via imperial channels. Flakes may have been removed before they were transferred to subordinate elites occupying the summit of Cerro Mejía. These subordinate elites may have also engaged in this type of "skimming" (Figure 4E). Elites charged with the supervision of commoners, such as those on Cerro Mejía, distributed these items to clients in small amounts, perhaps at events with feasting to mark occasions of calendrical import. Distribution of this kind may have fostered a patron-client relationship while avoiding

Table 2. Distribution of obsidian in the excavated houses on Cerro Mejía and Cerro Baúl.

Site/ Unit	CM Un3	CM Un4	CM Un5	CM Un6	CM Un17	CM Un18	CM Un19	CM Un20	CM Un118	CM Un136	CM Un145	CB Palace
M² excavated	34	54	12	13	51	35	51	29	313	82	228	445
Point Count	6	4	0	8	10	8	4	6	5	2	7	66
Point weight (g)	2.6	7.5	0	5.3	6.5	4.2	1.82	10.7	13.6	10.1	39.6	169.28
Others	51	10	98	5	170	49	737	111	498	136	131	267
Weight (g)	1.05	5.0	4.25	0.65	20.55	9.5	17.21	15.58	29.79	18.15	90.42	194.34
Total Number	57	14	98	18	180	64	741	116	503	138	138	333
Total Weight (g)	3.65	12.5	4.25	5.95	27.05	13.4	19.03	26.59	43.39	28.25	130.02	363.62
Density by Count	1.68	0.26	8.17	1.38	3.53	1.83	14.53	4.0	1.61	1.68	0.61	0.74
Density by Weight	0.11	0.23	0.35	0.46	0.53	0.38	0.37	0.92	0.14	0.34	0.57	0.81

the broader distribution of items such as decorated pottery that held a higher symbolic value (Nash 2019). Clients may have received nearly exhausted preforms, flakes, or small triangular points. Obsidian is ideal because it is consumed as it is used, although households could reserve obsidian for exchange or ritual offerings (Nash and deFrance 2019).

If this was the case, obsidian may be a better indicator of community connection with the Wari Empire than decorated pottery. The appearance or increase of obsidian resources in an area during the Wari era may stem from a state institution that distributed obsidian to create relations of reciprocity or indebted clients to state-sponsored patrons. I suspect there was more to the Wari political economy than feasting alone.

Context is important. Areas with an abundance of obsidian or communities without the technical skill may have followed different methods of reduction and use. However, in zones where few pieces of obsidian arrived before Wari incursion, such as Moquegua, the prevalence of obsidian debitage in households—and particularly the presence of the diagnostic Wari Laurel Leaf point—probably indicates participation in imperial channels of exchange or incorporation in the empire's political economy. Thus far, researchers have relied heavily on architecture and pottery to discern the presence and type of relationship between the Wari Empire and people living in different regions. In this chapter, I have proposed that obsidian and its sources, shapes, and types of production debris can offer another important line of evidence. This study demonstrates that lithic artifacts have a significant story to tell about the scope and strength of the Wari Empire and perhaps other complex societies as well.

References

Bard, Kathryn A.

1994   The Egyptian Predynastic: A Review of the Evidence. *Journal of Field Archaeology* 21(3): 264–288.

Bauer, Brian S.

1998   *The Sacred Landscape of the Inca: The Cusco Ceque System.* University of Texas, Austin.

Bencic, Catherine M.

2015   Lithic Production, Use and Deposition at the Wari Site of Conchopata in Ayacucho, Peru. PhD dissertation, Department of Anthropology, Binghamton University, Binghamton, New York.

Bergh, Susan

2012   *Wari: Lords of the Ancient Andes.* Thames and Hudson, New York.

Burger, Richard, Frank Asaro, Guido Salas, and Fred Stross

1998   The Chivay Obsidian Source and the Geological Origin of Titicaca Basin Type Obsidian Artifacts. *Andean Past* 5: 203–223.

Burger, Richard, and Michael Glascock

2000   Locating the Quispisisa Obsidian Sources in the Department of Ayacucho, Peru. *Latin American Antiquity* 11(3): 258–268.

Burger, Richard L., Karen L. Mohr Chávez, and Sergio J. Chávez

2000   Through the Glass Darkly: Prehispanic Obsidian Procurement and Exchange in Southern Peru and Northern Bolivia. *Journal of World Prehistory* 14(3): 267–362.

Druc, Isabelle, Miłosz Giersz, Maciej Kałaska, Rafał Siuda, Marcin Syczewski, Roberto Pimentel Nita, Julia M. Chyla, and Krzysztof Makowski

2020   Offerings for Wari Ancestors: Strategies of Ceramic Production and Distribution at Castillo de Huarmey, Peru. *Journal of Archaeological Science: Reports* 30: 102229.

Given, Michael

2004   *The Archaeology of the Colonized.* Routledge, London.

Isbell, William H.

1991   Huari Administration and the Orthogonal Cellular Architecture Horizon. In *Huari Administrative Structure: Prehistoric Monumental Architecture and State Government*, edited by William H. Isbell and Gordon F. McEwan, pp. 293–316. Dumbarton Oaks, Washington, DC.

Jennings, Justin, and Michael Glascock

2002   Description and Method of Exploitation of the Alca Obsidian Source, Peru. *Latin American Antiquity* 13(1): 107–117.

Kaplan, Jessica

2018   Obsidian Networks and Imperial Processes: Sourcing Obsidian from the Capital of the Wari Empire, Peru (AD 600–1000). PhD dissertation, University of California, Santa Barbara.

Klink, Cynthia, and Mark Aldenderfer

2005   A Projectile Point Chronology for the South-Central Andean Highlands. In *Advances in Titicaca Basin Archaeology – 1*, edited by Charles Stanish, Amanda Cohen, and Mark Aldenderfer, pp. 25–54. Cotsen Institute of Archaeology at UCLA.

Nash, Donna J.

2015   Evidencia de uniones matrimoniales entre las élites wari y tiwanaku de Cerro Baúl, Moquegua, Perú. In *El Horizonte Medio: Nuevos aportes para el sur de Perú, norte de Chile y Bolivia*, edited by Antti Korpisarri and Juan Chacama, pp. 177–200. Ediciones Universidad de Tarapacá, Arica, Chile.

2017   Clash of the Cosmologies: Vernacular vs. State Housing in the Wari Empire. In *Vernacular Architecture of the Pre-Columbian Americas*, edited by Christine Halperin and Lauren Schwartz, pp. 91–112. Routledge, London.

2018   Art and Elite Political Machinations in the Middle Horizon Andes. In *Images in Action: The Southern Andean Iconographic Series*, edited by William Isbell, Mauricio Uribe, Anne Tiballi, and Edward P. Zegarra, pp. 469–487. Cotsen Institute of Archaeology at University of California, Los Angeles.

2019   Craft Production as an Empowering Strategy in an Emerging Empire. *Journal of Anthropological Research* 75(3): 328–360.

2024   Precincts and Political Organization: Inferring Wari Integration from Site Configuration. In *District and Neighborhood Sociopolitical Integration in the Andes and Mesoamerica*, edited by Gabriela Cervantes and John Walden. Center for Comparative Archaeology, University of Pittsburgh, Pittsburgh, Pennsylvania.

Nash, Donna J., and Susan D. deFrance

2019   Plotting Abandonment: Excavating a Ritual Deposit at the Wari Site of Cerro Baúl. *Journal of Anthropological Archaeology* 53: 112–132.

Nash, Donna, and Patrick Ryan Williams

2005   Architecture and Power: Relations on the Wari-Tiwanaku Frontier. *Archaeological Papers of the American Anthropological Association* 14:151–174.

2009   Wari Political Organization on the Southern Periphery. In *Andean Civilization: A Tribute to Michael E. Moseley*, edited by Joyce Marcus and Patrick Ryan Williams, pp. 257–276. Cotsen Institute of Archaeology Monograph 63, Los

Angeles, California.

2021   As Wari Weakened: Ritual Transitions in the Terminal Middle Horizon of Moquegua, Peru. In *Rituals, Collapse, and Radical Transformation in Archaic States*, edited by Joanne M. A. Murphy, pp. 77–99. Routledge, New York.

Rademaker, Kurt

2006   Geoarchaeological Investigations of the Waynuna Site and the Alca Obsidian Source, Peru. MS thesis, Department of Quaternary and Climate Studies, University of Maine, Orono.

Shott, M. J.

1997   Stones and Shafts Redux: The Metric Discrimination of Chipped-stone Dart and Arrow Points. *American Antiquity* 62: 87–101.

Tripcevich, Nicholas

2007   Quarries, Caravans, and Routes to Complexity: Prehispanic Obsidian in the South-central Andes. University of California, Santa Barbara.

Tripcevich, Nicholas, and Daniel A. Contreras

2013   Archaeological Approaches to Obsidian Quarries: Investigations at the Quispisisa Source. In *Mining and Quarrying in the Ancient Andes*, pp. 23–44. Springer, New York.

Williams, P. Ryan, Laure Dussubieux, and Donna J. Nash

2012   Provenance of Peruvian Wari Obsidian: Comparing INAA, LA-ICP-MS, and Portable XRF. In *Obsidian and Ancient Manufactured Glasses*, edited by Ioannis Liritzis and Christopher M. Stevenson, pp. 75–85. University of New Mexico Press, Albuquerque.

Williams, Patrick Ryan, Donna J. Nash, Anita Cook, William Isbell, and Robert J. Speakman

2019   Wari Ceramic Production in the Heartland and Provinces. In *Ceramics of the Indigenous Cultures of South America: Production and Exchange*, edited by Michael D. Glascock, Hector Neff, and Kevin J. Vaughn, pp. 125–133. University of New Mexico Press, Albuquerque.

CHAPTER 8

A Revised Method for Confidently Sourcing Small Obsidian Artifacts

LUCAS R. M. JOHNSON, KYLE P. FREUND, M. KATHLEEN DAVIS, AND DARON DUKE

Lucas R. M. Johnson Far Western Anthropological Research Group, Inc., 1180 Center Point Dr., Suite 100, Henderson, NV 89074, USA (lucas@farwestern.com, corresponding author)

Kyle P. Freund Far Western Anthropological Research Group, Inc., 1180 Center Point Dr., Suite 100, Henderson, NV 89074, USA; Adjunct Assistant Professor at the University of Nevada, Las Vegas, USA

M. Kathleen Davis Far Western Anthropological Research Group, Inc., 2727 del Rio Pl., Davis, CA 95618, USA

Daron Duke Far Western Anthropological Research Group, Inc., 1180 Center Point Dr., Suite 100, Henderson, NV 89074, USA

Abstract

This chapter concerns archaeological obsidian sourcing using X-ray fluorescence (XRF) spectrometry and describes a revised method for confidently sourcing small artifacts that do not meet conventional assumptions of infinite thickness. A persistent problem for obsidian analysts is the distortion of parts per million (ppm) values for small and thin samples caused by uneven Compton normalization across analyzed elements due to increasing infinite thickness with atomic number. Despite these complications, XRF analysts continue to analyze small artifacts and make source assignments by transforming semi-quantitative photon count data or ppm values into element ratios or relative peak percentages, often expressing caveats with source assignments. While in certain contexts these transformations provide useful data for sourcing small and thin samples, we expand on these methods by creating 95% confidence regions that include small specimens of geologic source material and applying them to ternary diagrams. This allows for more confident source assignments of smaller artifacts from a wider range of archaeological contexts.

Introduction

This chapter adapts a method borrowed from the geological sciences for the sourcing of small obsidian artifacts that do not meet conventional assumptions of infinite thickness—i.e., that are not sufficiently thick to completely absorb instrumental X-rays and reflect them back to the detector (de Vries and Vrebos 2002; Davis 2011). This issue deserves attention because thin flakes and blades are a common occurrence in lithic assemblages, and they are often ignored in obsidian sourcing studies because they are assumed to be too small for analysis. This creates a bias in archaeological sampling strategies toward larger flakes and finished tools. As has been argued by Freund and colleagues (2022: 109), more attention to sampling strategies would allow source exploitation to be fully analyzed within the entire *chaîne opératoire* of regional lithic economies.

Obsidian sourcing studies have a long history in archaeological research and are employed to address a range of compelling research questions (Carter 2014; Eerkens et al. 2007; Freund 2013; Hughes 1986; Jackson 1989; Kuzmin et al. 2020; Shackley 1998). To identify the geological source of obsidian artifacts, numerous methods are available; and factors such as time, cost, size requirements, and destructiveness of the analysis must be considered. X-ray fluorescence (XRF) spectrometry is among the most common because it is both comparatively precise and accurate; it is also rapid, non-destructive, and relatively cheap when compared to other techniques (Tykot 2017).

The problem with small and thin obsidian specimens

Historically, XRF analysis has favored larger artifacts due to the understanding that thicker objects maximize fluorescence efficiency for mid-Z elements such as Rb through Nb. A minimum thickness of 4 mm is often cited, as it is the point at which all incoming X-rays are absorbed by the sample for Zr in an obsidian matrix (Ferguson 2012); however, in practice, 2.0 mm-thick samples can yield comparable part per million (ppm) values (Davis et al. 2011). As Davis and colleagues (1998, 2011) have shown, ppm values increase when objects are at, or thinner than, 1.7 mm in thickness (Figure 1). These skewed ppm measurements are typically caused by normalizing to the Compton scatter peak or by some other physics-based parameter of the calibration, such as Fundamental Parameters with standards (see Johnson et al. 2021).

To demonstrate these distortions, Figure 2 shows analyses of geological obsidian samples of varied thickness from Glass Mountain (California). The thicker sample shows higher element peaks and Compton scatter as well as a higher count rate per second—or *valid count*, as it is known in the Bruker S1PXRF software. Calibrated ppm values for the three specimens (see Johnson et al. 2021 for analytical conditions) vary widely in at least

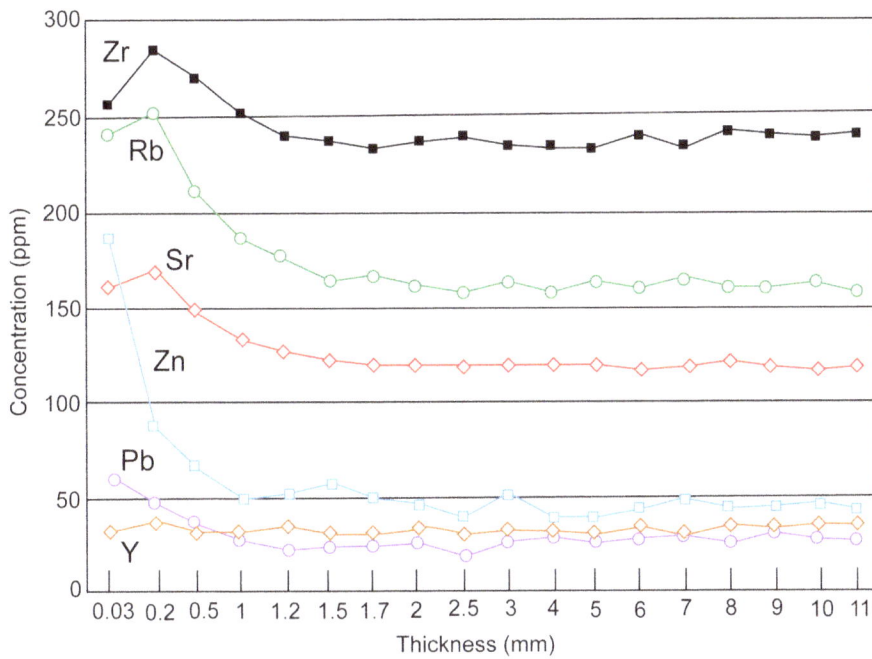

Figure 1. Distortion in ppm by sample thickness using USGS sample RGM-1. Redrawn from Davis et al. 2011.

three elements, resulting in either overpredicted or underpredicted values as sample thickness decreases. Consistent with the findings of Davis and colleagues (2011), described above, these tend to be systematic, depending on the concentration of a given element. In other words, ppm values for small and thin artifacts tend to trail away from those of infinite thickness, showing increasing ppm values with diminishing thickness (Glascock 2020: 42; see Figure 2).

To better understand the relationship between artifact size and counts, a discussion of raw and valid counts is necessary. Fluorescence efficiency can be understood by determining the count rate, or photons per second, recorded by an instrument's detector as X-rays return to the instrument. For Bruker products manufactured prior to about

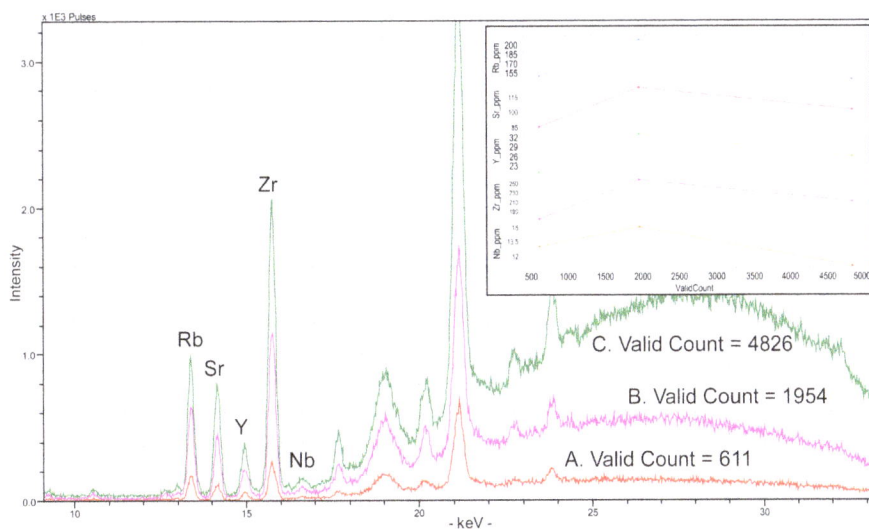

Figure 2. Glass Mountain obsidian of varying thickness showing keV energy by intensity. Inset shows decreasing valid count and ppm for the three samples. Sample A measures 9.7 x 12.4 x 1.4 mm; Sample B measures 12.9 x 9.2 x 2.5 mm; Sample C measures 28.0 x 19.6 x 16.5 mm.

2018, this is referred to as *raw count* and is a software-measure of ionized particles from the analyte and secondary scattering. The "noise" from the instrument is either filtered out or subtracted via the software, thus producing a *valid count*. The difference between the *raw count* and *valid count* is typically about 200 photons per second. Valid count is correlated with sample mass. For example, those samples with thicknesses of 4 mm or greater typically return valid counts between about 3,000 and 5,100 counts per second, while those samples between 0.5 and 2 mm typically return valid counts between 540 and 1,500. Thus, valid count can be used as an approximate index for sample size or shape.

An understanding of the correlation between sample size and valid count or count rate is useful because one must consider the size and shape of the geological samples used to make source assignments. Indeed, most geological source library samples consist of larger, thicker samples (see Ferguson 2012; Glascock 1998). This means that those artifacts with geometry (i.e., size and shape) similar to geologic source materials will yield equivalent count rates and thus align statistically with calculated ppm values. Ppm values for smaller and thinner artifacts, by contrast, are typically skewed in the manner described above.

The following example illustrates these issues in further depth. From a recent project in the Western Great Basin, more than 1,000 artifacts were analyzed. Their valid count rates per second ranged from 93 to 3,809 (mean = 2,036, Std = 817, CoV = 40%). This distribution alone suggests that ppm will be highly skewed when compared to a source library composed of large and thick specimens. The range in valid count was also recorded for 400 Bodie Hills (California) geologic source specimens, part of a simulated reduction exercise that produced more than 1,200 flakes of the material, all of which were measured for length, width, and thickness and analyzed with the same ED-pXRF instrument. This experimental study reduced cobbles to produce a variety of core reduction and retouch flakes as well as smaller shatter. Results from the analysis of the 400 Bodie Hills samples show a noticeable skew in most ppm measurements when compared to results for thicker samples (Table 1).

Conventional methods used to source small artifacts

There are several ways to assign artifacts to a source (see Shackley 2011 for a review), but the most common is to compare the elemental signatures of artifacts with that of geological source material analyzed with the same instrumentation—in a sense, "fingerprinting" a sample to known geological outcrops. As discussed above, this is inherently problematic for small artifacts because ppm values for specimens thinner than approximately 2 mm may be significantly distorted: but source assignments can still be made.

Since XRF measurement errors on small artifacts are systematic, element ratios can be used to make source assignments. Displaying elemental peak percentages in ternary diagrams is one way to avoid direct comparisons of ppm measurements (Hughes 2010; Panich 2016). While these techniques can be applied to certain archaeological contexts, depending on the complexity of the regional geochemistry, source assignments cannot be made with the same level of confidence as larger artifacts due to systematic offset of small samples and their appearance as outliers relative to 95% confidence ellipses (i.e., two standard deviations) (see Glascock 2020: 44, Figure 5).

Table 1. Ppm statistics by sample thickness dimension for 400 Bodie Hills specimens created through experimental bipolar reduction.

Element (ppm)	n=36, < 1mm	n=142, 1–2mm	n=161, > 2–4mm	n=61, > 4mm	n=400
Rb					
ppm min-max	151–288	153–268	159–251	144–221	144–288
Mean	213	219	200	178	204
Std	33	21	19	18	25
CoV	15%	9%	9%	10%	12%
Sr					
ppm min-max	91–151	84–151	75–143	75–120	75–151
Mean	115	109	100	94	104
Std	15	14	11	10	14
CoV	13%	13%	11%	10%	13%
Y					
ppm min-max	7–30	7–29	8–29	7–26	7–30
Mean	16	14	14	15	14
Std	6	6	5	5	5
CoV	42%	42%	37%	35%	39%
Zr					
ppm min-max	96–173	88–177	93–167	90–155	88–177
Mean	130	123	117	115	120
Std	20	21	18	18	20
CoV	15%	17%	15%	15%	16%
Nb					
ppm min-max	7–20	12–21	12–20	11–19	7–21
Mean	14	16	15	14	15
Std	2	2	2	2	2
CoV	18%	11%	10%	12%	13%

Use of element ratios

The use of element ratios to source artifacts has been applied to a range of archaeological contexts, including the western US (Hughes 2007), the central Mediterranean (e.g., Freund 2014), the eastern Mediterranean (e.g., Frahm 2016), Mesoamerica (Stroth et al. 2019), and numerous others.

For example, Hughes (2007) describes the use of peak intensity ratios to source small artifacts. His ratios of Fe/Mn, Rb/Sr, Zr/Y, Y/Nb, Zr/Nb, and Sr/Y are applied to sources in California, Nevada, Oregon, Utah, Idaho, and Arizona. Similarly, Frahm (2016) outlines the use of ppm ratios in the eastern Mediterranean that are provided by their predictive strengths: Fe/Mn, Zr/Nb, Zr/Sr, Sr/Nb, Rb/Nb, Rb/Zr, Rb/Sr. The major difference between Hughes (2007) and Frahm (2016) is that Hughes assesses bivariate plots and confidence ellipses whereas Frahm applies a discriminate function analysis (DFA). The DFA approach assumes distinct separation between sources (see Glascock 1998, 2020) and requires analysis of all known sources from a given region. The use of ppm ratios is required in DFA because they are based upon independent variables (e.g., calibrated ppm) rather than instrument-specific ratios of peak counts.

Element peak percentages in a ternary diagram

A ternary diagram, often referred to as the "simplex" in geological sciences, is a useful way to visualize the relative contributions of three variables when they are summed to 100%. Jack and Heizer (1968) were the first to display relative peak percentages of obsidian artifacts using ternary diagrams with rough outlines of various sources in Mesoamerica (Figure 3). As Hughes (1986: 50; 2010) explains, this method simply adds three elements and divides by the sum, each having a value between 0 and 100% and plotted on its own axis. This three-dimensional method plotted on a two-dimensional image displays the relative contribution of element counts, not ppm. Subsequent iterations of this plot are published in Hughes (1986), Jackson (1989), and Shackley (1988).

Hughes (2010) offers a more refined use of the ternary diagram, and he again targets the elements Zr, Sr, and Rb because they are adjacent on the periodic table and respond similarly to variations in sample thickness. In this much refined image (Hughes 2010: Figure 3), small artifacts are plotted with dashed ovals representing the range of variation of a given source. These ovals are inferred from extensive knowledge of how particular sources vary with thickness and diameter. They are not statistically defined, however, and thus are difficult to reproduce.

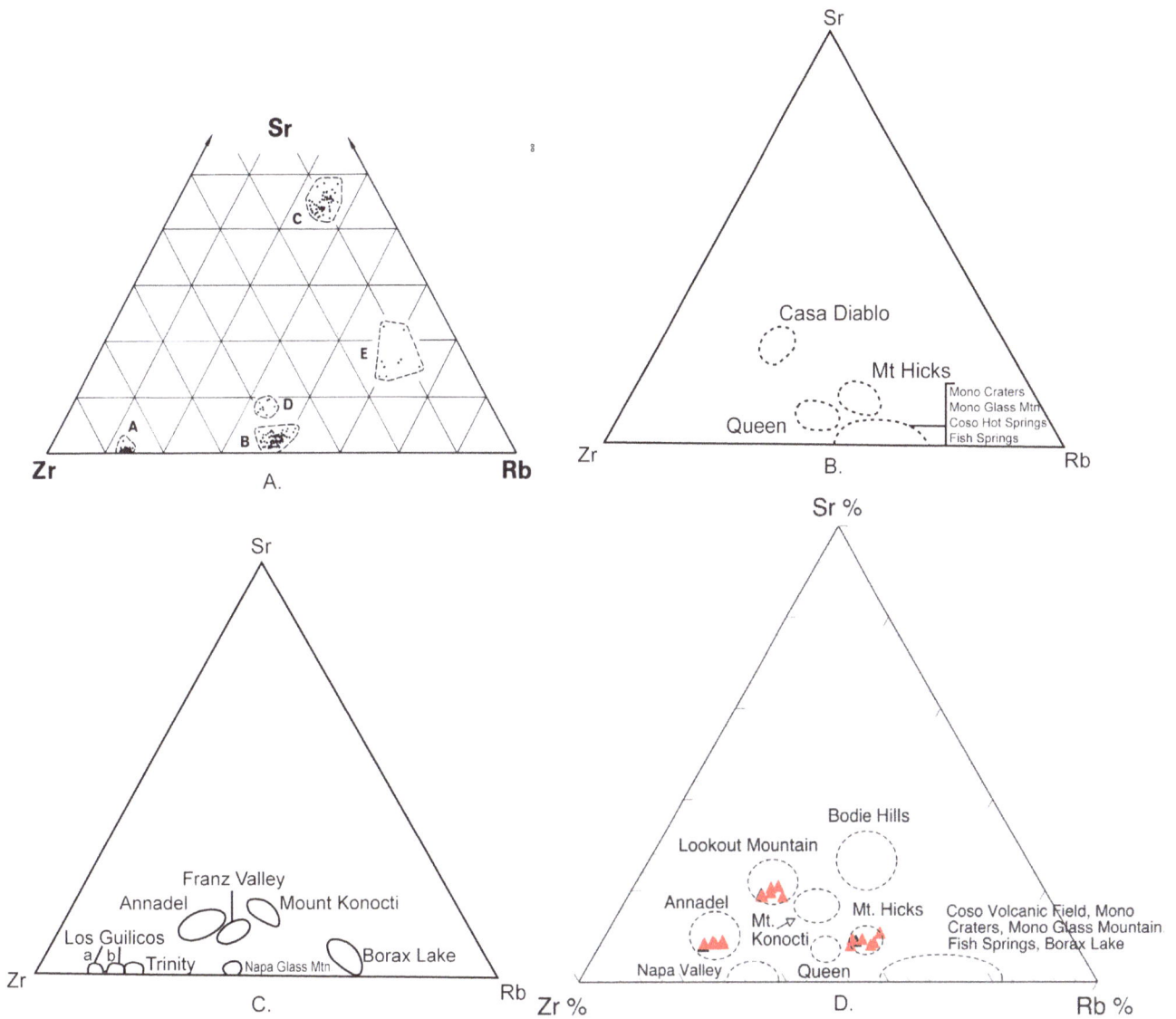

Figure 3. Various ternary diagrams. A – taken from Jack and Heizer (1968); B – redrawn from Hughes (1986); and C – redrawn from Jackson (1989); D – taken from Hughes (2010).

Expanding a method for sourcing small obsidian artifacts

Many statistical programs offer the ability to plot data in ternary diagrams (see Panich 2016), but more robust statistical methods that predict ranges of variation require additional steps and expanded inputs. In geology, ternary diagrams are often used to plot the relative percentage of feldspar with other mineralogical components (Weltje 2002; Hamilton and Ferry 2018), and confidence regions are often applied to the ternary diagrams to aid in classification of rock types. There are several methods for their calculation (Hamilton and Ferry 2018; Medak and Cressie 1991; Tolosana-Delgado and van den Boogaart 2011; Watson and Nguyen 1985; Weltje 2002). Most use a modified Mahalanobis Distance statistic, or D^2, and require extensive calculations to compute depending on how many variables are included.

To calculate confidence regions in ternary diagrams for use in archaeological obsidian sourcing, the R Statistical Program can be used with the *ggtern* package and other dependent applications (e.g., *ggplot2*). The code, written by Hamilton and Ferry (2008), is straightforward and can be expanded upon for visualization or additional groupings (Figure 4). While specific applications of the D^2 statistic have been applied to establish outliers in obsidian sourcing (see Glascock 1998), here it is modified to calculate a non-Euclidean distance from a central tendency (Hamilton and Ferry 2018). This non-Euclidean distance is then used to calculate a modified Z-score that establishes how many standard deviations a sample is from that central point in three dimensions. Non-Euclidean, in this instance, refers to the fact that the data ranges from 0 to 100% and is thus not continuous (i.e., $0 - n$). Confidence regions, like ellipses, can be estimated at the 50, 90, 95, or 99% confidence level, where 95% estimations indicate those samples that are no more than two standard deviations from the mean.

Figure 4. R *ggtern* script credited to Hamilton; adapted for L. R. M. Johnson from Hamilton and Ferry (2018) and R Core Team (2021).

```
+ library(ggtern)
+ #Root Directory
+ setwd("C:/Users/…")
+ #Load Data
+ df = read.csv("./… .csv",header=TRUE,stringsAsFactors = FALSE)
+ #Rename Empty Sources
+ df$SourceName[which(as.character(df$Source.Final) == "")] = "<Unknown>"
+ #Base Plot
+ base = ggtern(df,aes(Zr.,Sr.,Rb.)) +
+  theme_bw() +
+ stat_confidence_tern(geom='polygon',
+              aes(group=SourceName,color=SourceName,fill=SourceName),
+              alpha=0.95,breaks=c(0.99,0.95,0.5)) +
+ geom_point(aes(color=SourceName,fill=SourceName),size=0.5,shape=21,color='black') +
+ scale_size_identity() +
+ guides(color = guide_legend(override.aes = list(size = 3))) +
+ labs(title = "Ternary Plot with Confidence Regions") +
+ base
+ pdf("… .pdf",width=8,height=6)
+  print(base)
+ dev.off()
```

The basic data inputs for confidence region prediction include known source groups and their relative element percentages (see above). Unlike geological studies of relative mineral concentration, application of the method to obsidian sourcing requires a thorough evaluation of geochemical groups (see Glascock 1998). It is important to note that this method is not intended as a substitute for the use of ratios described earlier (e.g., Sr/Zr), but to supplement and expand on the earlier work of Jack and Heizer (1968) and Hughes (2010), enabling a reproducible procedure for calculating statistical confidence regions within ternary diagrams. In some instances, dozens of sources can be plotted to reflect a geochemical resource map, although sources with similar geochemistry will overlap.

To apply confidence statistics to a ternary diagram that reflects both small and large artifact sizes, the problem becomes how to replicate this variation in a source library, remembering that source libraries are conventionally comprised of infinitely thick samples. Figure 5 shows 24 obsidian sources from the western US plotted on a ternary diagram with only infinitely thick source samples. One can see the relatively tight region depicted for each source. These 95% confidence regions, unlike confidence ellipses derived from two variables, are based upon relative percentages of three variables (Rb, Sr, and Zr) and are calculated using a Mahalanobis Distance or D^2 method in R.

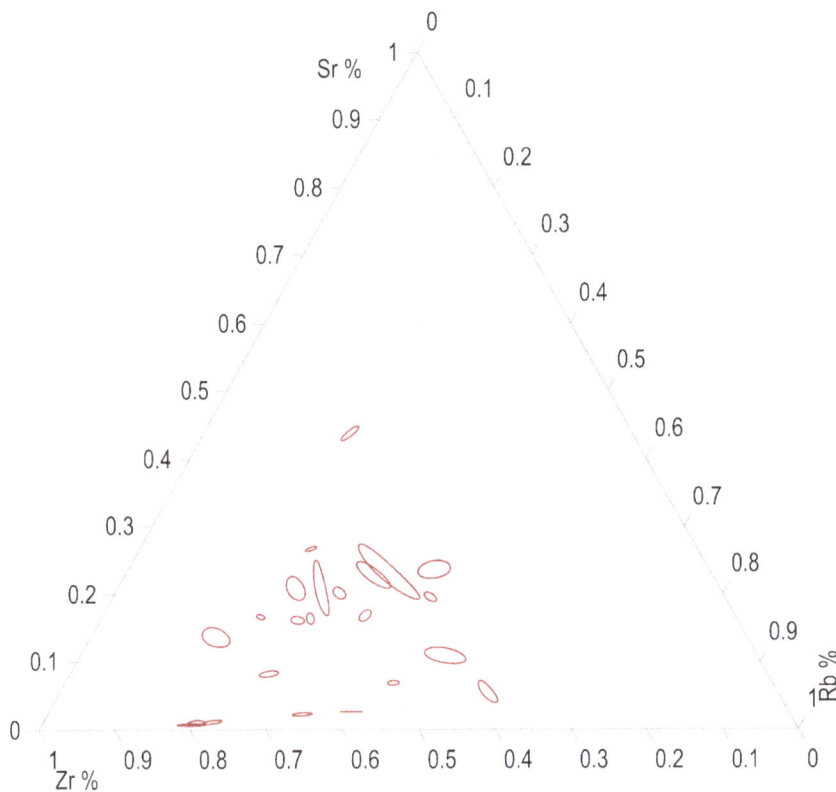

Figure 5. Example of ternary diagram and 95% confidence regions for 24 obsidian sources where each sample is infinitely thick (≥ 4 mm).

To expand the 95% confidence region so that it encompasses the variability introduced by small artifacts, other sizes of source material are required. To this end, we expanded the reference material to include small percussion or pressure flakes made from the same nodules as the large (infinitely thick) flakes. With this addition, the size of the confidence regions increases significantly (Figure 6), creating a bimodal distribution in valid count. The number of small/large flake pairs per source ranges from a low of 4 to a high of 60, with an average of about 17 for both. Some reference samples were unavailable for small flake production; consequently, there are more large than small flakes in the reference material for some sources (see Figure 6B). The majority, however, consist of small/large flake pairs.

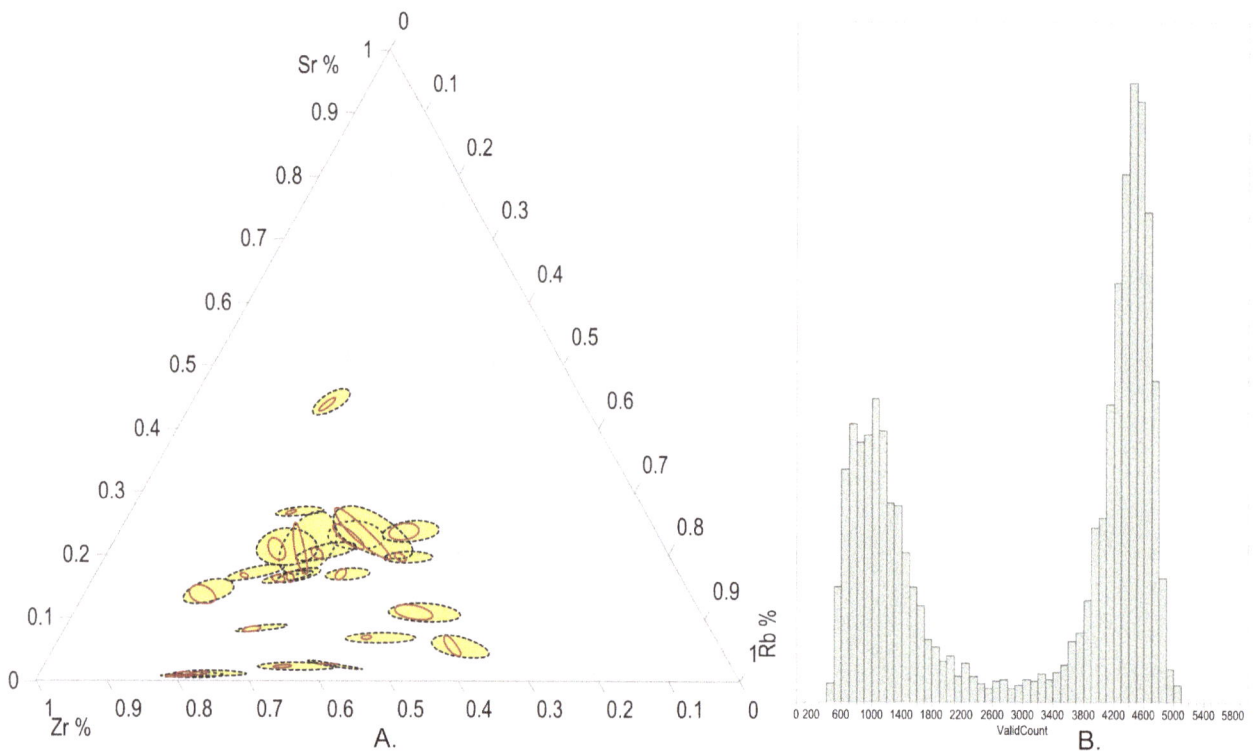

Figure 6. A – ternary diagram with both small/thin and large/infinitely thick source samples represented by 95% confidence regions. B – distribution of valid count for these sources. Statistics for this source library sample: n = 4,752; valid count range = 453–5,058; mean = 3,192; Std = 1,526; CoV = 47%.

Case studies from California and the Western Great Basin

To demonstrate the utility of sourcing both large and small obsidian artifacts using ternary plots with 95% confidence regions, two case studies from the western US are presented.

Case study 1

Approximately 500 artifacts were analyzed in north-central California along the Feather River as part of a large water management project (Figure 7). The project area is located at the confluence of multiple Indigenous groups that resided in the Central Valley as well as in the uplands of northwestern Nevada. Multiple habitation sites were investigated where tool retouch occurred, thus most samples were small pressure flakes and other retouch flakes. Figure 8 shows that many of the small specimens were made from California obsidian sources, as well as those near the modern-day state border (e.g., Buffalo Hills and Massacre Lake).

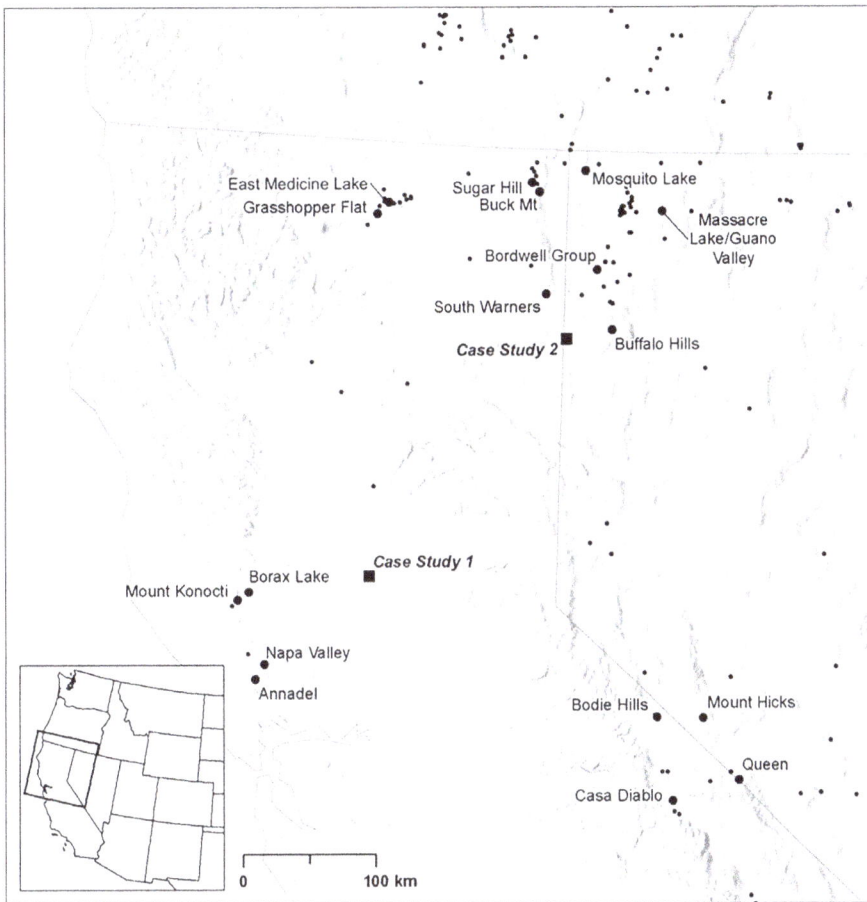

Figure 7. Case study locations with major obsidian sources labeled.

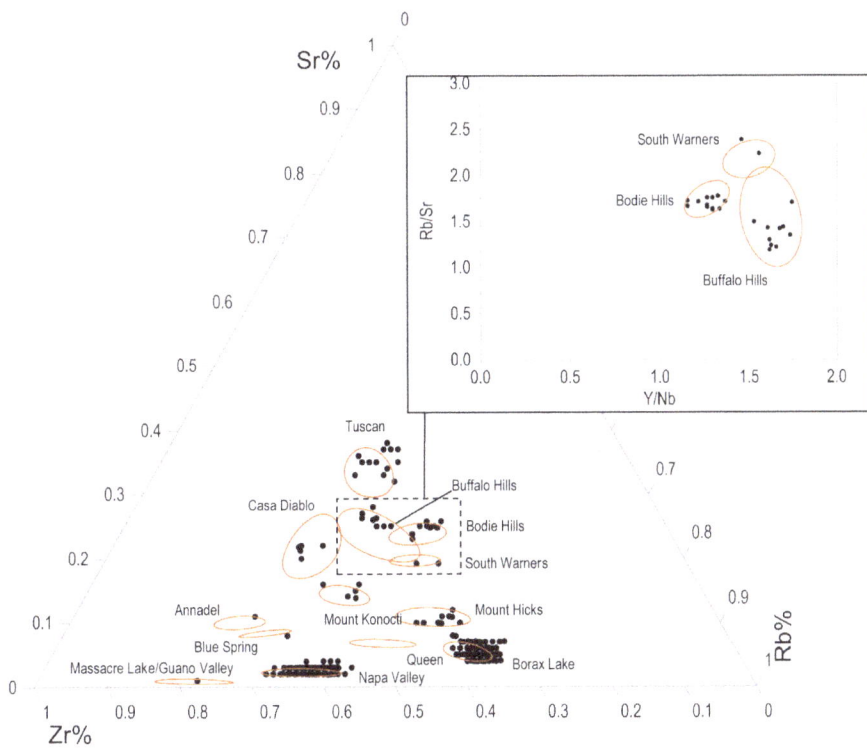

Figure 8. Case study 1 data. Ternary plot of relative peak percentages for elements Rb, Sr, and Zr with 95% confidence regions for sources derived from large and small samples. Confidence regions are calculated using the ggtern package in R (Hamilton 2018).

Case study 2

Another substantial sourcing project was conducted in northwestern Nevada during investigations to restore freshwater springs (Figure 7). Again, multiple obsidian *source regions* were identified in the artifact sample (Figure 9). Here, two major clusters of artifacts are shown. In the center are sources local to the archaeological sites (e.g., Buffalo Hills) as well as sources imported from northeast California (e.g., South Warners and Buck Mountain). In the lower left of the plot are three major northwestern Nevada sources (e.g., the Bordwell Group and Massacre Lake). While these sources overlap in the ternary plot and are plotted as a single Northwest Nevada Group, the addition of bivariate plots shows a clear separation. In this example, the use of Fe/Mn peak intensity ratio is necessary for the separation. In these peak intensity ratio bivariate plots, 95% confidence ellipses are shown for both large and small source samples combined, appearing as a dashed-line ellipse.

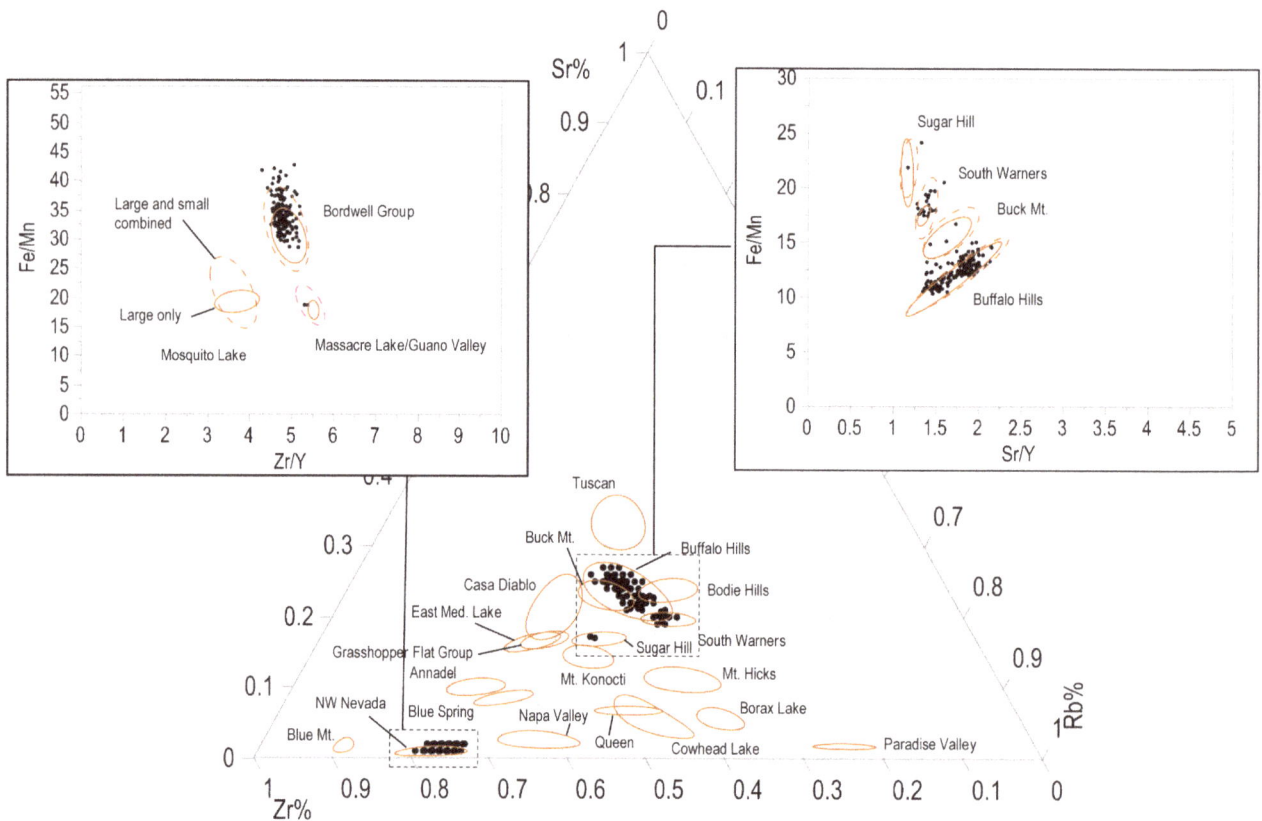

Figure 9. Case study 2 data. Ternary plot of relative peak percentages for elements Rb, Sr, and Zr, showing 95% confidence regions for sources likely to occur in the study area. Artifacts appear in black. Confidence regions are derived from large and small source specimens and are calculated and drawn with the ggtern package in R (Hamilton 2018).

Discussion and conclusions

The sourcing of small artifacts from archaeological assemblages using XRF has always been problematic because ppm values become skewed when specimens are thinner than 1.7 mm (Davis et al. 1998). This issue can be sidestepped by transforming semi-quantitative photon count data or ppm values into element ratios or relative peak percentages.

Considering these issues, this chapter outlines a refined method for confidently sourcing both large and small obsidian artifacts by applying 95% confidence regions to ternary diagrams to encapsulate a wider range of expected variation in geological sources regardless of specimen size, thus making source assignments easier to determine. Borrowing from methods used in the geological sciences and only recently applied to archaeological obsidian sourcing, this chapter specifically discusses how to use the R Statistical Program to calculate the Mahalanobis Distance statistic and, by extension, confidence regions in ternary diagrams.

As with any method, some caveats apply. Chief among them is that the comparison of source confidence regions created with small flakes assumes that the small flakes chosen from one source are morphologically equivalent to the small flakes from another source. This is a non-trivial problem for the analyst choosing the flakes and bound to be an approximation at best. A greater number and more complete range of flake sizes than we have used in this study (that is, more than just large flakes and small percussion/pressure flakes) would go some distance toward correcting this source of error. It will also be obvious that some chemically similar sources will become impossible to distinguish when small flake distortions are introduced (for example, the Coso subsources in southeast California). In these instances, however, a more complete range of flake sizes in the reference collection could help refine the threshold for discrimination (see Hughes 2010). This issue, along with a more detailed evaluation of the accuracy of relative peak percentages as they are applied to small artifact analysis, will be considered in a forthcoming study by the authors.

Determining the origin of obsidian artifacts is a powerful tool that can be used to address a wide range of relevant archaeological issues. However, due attention must be paid to sampling strategies. When the entire complement of lithic artifacts found within archaeological assemblages can be analyzed, it becomes possible to fully reconstruct lithic reduction sequences and the movement of raw materials across space. Importantly, we argue that the use of confidence statistics in ternary diagrams is but one tool for obsidian sourcing and should be used in conjunction with other methods of assignment depending on regional geochemistry.

Acknowledgments

While many months were spent researching how to reproduce confidence ellipses used to characterize unknown obsidian samples as seen in the cited literature, some dating back to the late 1960s, the credit for the method using the *ggtern* package in R goes to Nicholas E. Hamilton and his collaborator Michael Ferry. Nicholas was eager enough to reply to a stranger from across the globe with a very helpful technique that allows for the reproduction of our plots by anyone so long as they follow the steps described. Thanks to Amtrak for use of their Wi-Fi and working space on my commute between Berkeley and Davis. Nicholas was contacted before the writing of this chapter and was informed about our intention to publish a portion of the code he shared with me in early 2018. We wish to thank him for sharing the open-source code he developed and that we later modified slightly to suit our aesthetic needs. It was through our understanding of XRF physics on small and thin samples that we applied the code to sourcing small obsidian artifacts. With that, I also wish to thank Kasey O'Horo for her insights in trying to troubleshoot the issues present above. She was the person that encouraged us to look beyond infinitely thick samples. Additionally, we are grateful to Alexxandria Martinez for her continued service at Far Western XRF Lab in Henderson, NV. Her steadfast work ensures quality measurements are taken during each XRF scan. Similarly, we wish to thank Laura Harold and all the Far Western Principals for their continued support in growing the Far Western XRF Lab. The lab, split between two branch offices, is truly a collaborative endeavor.

References

Carter, T.
 2014   The contribution of obsidian characterization studies to early prehistoric archaeology. In *Lithic Raw Material Exploitation and Circulation in Prehistory*, edited by M. Yamada and A. Ono, pp. 23–33. ERAUL, Liège.

Davis, M. K., Jackson, T. L., Shackley, M. S., Teague, T., and J. Hampel
 1998   Factors affecting the energy-dispersive X-ray fluorescence (EDXRF) analysis of archaeological obsidian. In *Archaeological Obsidian Studies: Method and Theory*, edited by M. S. Shackley, pp. 159–180. Springer/Plenum, New York.
 2011   Factors affecting the energy-dispersive X-ray fluorescence (EDXRF) analysis of archaeological obsidian. In *X-Ray Fluorescence Spectrometry (XRF) in Geoarchaeology*, edited by M. S. Shackley, pp. 45–64. Springer, New York.

de Vries, J. L., and B. A. R. Vrebos
 2002   Quantification of infinitely thick specimens by XRF analysis. In *Handbook of X-Ray Spectrometry*, edited by R. E. Van Grieken and A. A. Markowicz, pp. 341–405. Marcel Dekker, New York.

Eerkens, J. W., J. R. Ferguson, M. D. Glascock, C. E. Skinner, and S. A. Waechter
 2007   Reduction strategies and geochemical characterization of lithic assemblages: a comparison of three case studies from Western North America. *American Antiquity* 72: 585–597.

Ferguson, J. R.
 2012   X-ray fluorescence of obsidian: approaches to calibration and the analysis of small samples. In *Studies in Archaeological Sciences: Handheld XRF for Art and Archaeology*, edited by A. N. Shugar and J. L. Mass, pp. 401–422. Leuven University Press, Leuven.

Frahm, E.
 2016   Can I get chips with that? Sourcing small obsidian artifacts down to microdebitage scales with portable XRF. *Journal of Archaeological Sciences Reports* 9: 448–467.

Freund, K. P.
 2013   An assessment of the current applications and future directions of obsidian sourcing in archaeological research. *Archaeometry* 55(5): 779–793.
 2014   Obsidian exploitation in Chalcolithic Sardinia: a view from Bingia ʻe Monti. *Journal of Archaeological Science* 41: 242–250.

Freund, K. P., L. R. M. Johnson, D. Duke, and D. Craig Young

2022　The character and use of the Ferguson Wash obsidian source in eastern Great Basin prehistory. *Journal of California and Great Basin Anthropology* 42(1): 101–115.

Glascock, M. D.

2020　A systematic approach to geochemical sourcing of obsidian artifacts. *Scientific Culture* 6(2): 35–47.

Glascock, M. D., G. E. Braswell, and R. H. Cobean

1998　A Systematic Approach to Obsidian Source Characterization. In *Archaeological Obsidian Studies: Method and Theory: Advances in Archaeological and Museum Science*, Vol. 3., edited by M. S. Shackley, pp. 15–66. Plenum Press, New York.

Hamilton, N. E.

2018　*ggtern: An Extension to 'ggplot2', for the Creation of Ternary Diagrams.* R Package Version 2.2.2. https://CRAN.R-project.org/package=ggtern, accessed August 1, 2022.

Hamilton, N. E., and M. Ferry

2018　ggtern: Ternary Diagrams Using ggplot2. *Journal of Statistical Software* 87: 1–17.

Hughes, R. E.

1987　*Diachronic Variability in Obsidian Procurement Patterns in Northeastern California and Southcentral Oregon, Publications in Anthropology*, Vol. 17. University of California Press, Berkeley.

2007　The Geologic Sources for Obsidian Artifacts from Minnesota Archaeological Sites. *The Minnesota Archaeologist* 66: 53–68.

2010　Determining the Geologic Provenance of Tiny Obsidian Flakes in Archaeology Using Nondestructive EDXRF. *American Laboratory* 42(7): 27–31.

Jack, R. N., and R. F. Heizer

1968　"Finger-printing" of some Mesoamerican obsidian artifacts. *Contributions of the University of California Archaeological Facility* 5: 81–100.

Jackson, T. L.

1989　Late Prehistoric Obsidian Production and Exchange in the North Coast Ranges, California. *Contributions of the University of California Archaeological Facility* 48: 79–84.

Jenkins, R.

1999　*X-Ray Fluorescence Spectrometry, Second Edition.* Monographs in Chemical Analysis 152. John Wiley & Sons, New York.

Johnson, L. R. M., J. R. Ferguson, K. P. Freund, L. Drake, and D. Duke
    2021   Evaluating Obsidian Calibration Sets with Portable X-Ray Fluorescence (ED-XRF) instruments. *Journal of Archaeological Science: Reports* 39: 103126.

Kuzmin, Y. V., C. Oppenheimer, and C. Renfrew
    2020   Global perspectives on obsidian studies in archaeology. *Quaternary International* 542: 41–53.

Medak, F., and N. Cressie
    1991   Confidence regions in ternary diagrams based on the power-divergence statistics. *Mathematical Geology* 23(8): 1045–1057.

Panich, L. M.
    2016   Beyond the colonial curtain: Investigating indigenous use of obsidian in Spanish California through the pXRF analysis of artifacts from Mission Santa Clara. *Journal of Archaeological Sciences: Reports* 5: 521–530.

Pollard, A. M., C. M. Batt, B. Stern, and M. M. Young
    2007   *Analytical Chemistry in Archaeology.* Cambridge University Press, Cambridge.

R Core Team
    2021   *R: A language and environment for statistical computing.* R Foundation for Statistical Computing, Vienna, Austria. http://www.R-project.org/, accessed August 1, 2022.

Shackley, M. S.
    1988   Sources of archaeological obsidian in the Southwest: An archaeological, petrological, and geochemical study. *American Antiquity* 53(4): 752–772.

Shackley, M. S. (editor)
    1998   *Archaeological Obsidian Studies: Method and Theory.* Advances in Archaeological and Museum Studies, Vol. 3. Plenum, New York.
    2011   *X-Ray fluorescence spectrometry (XRF) in archaeology.* Springer, New York.

Stroth, L., R. Otto, J. T. Daniels, and G. E. Braswell
    2019   Statistical artifacts: Critical approaches to the analysis of obsidian artifacts by portable X-ray fluorescence. *Journal of Archaeological Sciences: Reports* 24: 738–747.

Tolosana-Delgado, R., and K. G. van den Boogaart
    2011   Linear models with compositions in R. In *Compositional Data Analysis: Theory and Applications*, edited by V. Pawlowsky-Glahn and A. Buccianti, pp. 356–371. Wiley, New York.

Tykot, R. H.

2017   A decade of portable (hand-held) X-ray fluorescence spectrometer analysis of obsidian in the Mediterranean: many advantages and few limitations. *MRS Advances* 2(33–34): 1769–1784.

Watson, G. S., and H. Nguyen

1985   A confidence region in a ternary diagram from point counts. *Mathematical Geology* 17(2): 209–213.

Weltje, G. J.

2002   Quantitative analysis of detrital modes: statistically rigorous confidence regions in ternary diagrams and their use in sedimentary petrology. *Earth-Science Reviews* 57 (3–4): 211–253.

CHAPTER 9

Chlorine to Sodium Ratio as an Empirical Geochemical Estimator of Obsidian Aging

FRANCO FORESTA MARTIN, ENRICO MASSARO, PAOLA DONATO, AND
ROBERT H. TYKOT

Franco Foresta Martin Istituto Nazionale di Geofisica e Vulcanologia, Sezione di Palermo, and Laboratorio Museo di Scienze della Terra Isola di Ustica, Palermo, Italy
Enrico Massaro IAPS, Istituto Nazionale di Astrofisica, Rome, Italy
Paola Donato DiBEST, Università della Calabria, Cosenza, Italy
Robert H. Tykot University of South Florida, Tampa, Florida, USA

Abstract

This work is an extension of previous research in which we successfully tested the effectiveness of chlorine (Cl) versus sodium (Na) diagrams to unambiguously identify the provenance of obsidian artifacts attributable to Italian outcrops widely exploited during prehistory, i.e., Lipari, Pantelleria, Palmarola, and Monte Arci. In this further work, we found that the ratio R = Cl/Na decreases with the age of emplacement (t) of the obsidian outcrops and can be well described by the inverse law: $R(t) = 1/(A+Bt)$. Obsidian samples were analyzed both by electron probe micro-analyzer (EPMA) and by Instrumental Neutron Activation Analysis (INAA). Radiometric dates have been obtained both from the literature and through $^{40}Ar/^{39}Ar$ dating. Data processing confirmed the time evolution of the Cl/Na ratio through the same equation, indicating that it can be used as an empirical estimator of the obsidian formation age. A best-fit analysis of the collected data gives the relation $t = 149.3(1/R - 12.68)$ in units of ka. To verify the validity of this relation for non-Italian obsidians, we applied it to Sierra de Las Navajas (State of Hidalgo, Mexico) obsidians, estimating an age of 1.75 Ma, in agreement with the upper limit of 1.8–2 Ma generally accepted for these rocks. We propose that the Cl/Na ratio changes with time because of the differential loss of chlorine and sodium as a consequence of the micro-fracturing of the obsidian glass after the emplacement. If future tests on different obsidians can confirm the validity of this approach, a rough estimate of the age of emplacement of the obsidian outcrops could be derived from their geochemical compositions.

Introduction

The background

This chapter represents the further development of a study that aims to verify to what extent some analytical data relating to the concentration of chlorine (Cl) in obsidian can be used to obtain information of geologic and archaeometric interest on these volcanic glasses. In a previous study, Foresta Martin and colleagues (2020) proposed a new geochemical method to determine the provenance of the archaeological obsidian artifacts collected in the area of the Central Mediterranean, and attributable to the four Italian sources exploited during prehistory: Lipari (Aeolian Islands, Sicily), Pantelleria (Sicily), Palmarola (Pontine Islands, Latium), and Monte Arci (Sardinia) (Figure 1; see Freund 2018).

The method is based on measurements of the concentration of Cl that, although a minor element in volcanic glasses, nevertheless exhibit a good quantitative differentiation between the Italian obsidian sources (Foresta Martin et al. 2020) (Table 1).

The diagnostic relevance of Cl for obsidian studies can be better appreciated if we consider that in silica-rich magmas it acts as an incompatible element that tends to concentrate in melts rather than in the crystal lattices (Bonifacie et al. 2008). Furthermore, Cl solubility rises with the increasing content of network modifying cations, especially Na, K, Mg, Ca, and Fe, so that its concentration grows in more evolved magmas, such as trachytes and rhyolites, from which obsidian is generated, reaching significantly higher values (up to 0.50 wt%) than the average of igneous melts (0.02 wt%) (Lowenstern 1994; Carroll and Webster 2018 and references therein). In some peralkaline[1] obsidian, such as the pantellerites, Cl can even exceed 1 wt% (Lanzo et al. 2013). Sodium (Na) is another element that exhibits some quantitative differentiation between the Italian obsidian sources (Le Bourdonnec et al. 2006, 2010; Tykot 2002), although, taken on its own, it does not always provide unambiguous diagnoses of provenance.

After verifying that in the Central Mediterranean obsidian Cl and Na are positively correlated, Foresta Martin and colleagues (2020) tested the effectiveness of a Cl vs. Na_2O scatter plot to unambiguously discriminate the four Italian obsidian sources, carrying out measurements of the concentration of these two elements both in geological and archaeological samples, by non-destructive or minimally destructive and commonly used analytical procedures such as Scanning Electron Microscopy / Energy Dispersive X-ray Spectroscopy (SEM-EDS) and electron probe micro analysis (EPMA). The results were conclusive, as shown by the Cl vs. Na_2O scatter plot (Figure 2), where both geological and archaeological obsidian samples from several Italian outcrops are grouped over distinct areas, without overlapping (Foresta Martin et al. 2020).

Table 1. Mean Cl concentrations in wt%. In brackets are the number of analyzed (EPMA) samples. LIP = Lipari (GAB = Gabellotto; CD = Canneto Dentro; MGU = Monte Guardia). PANT = Pantelleria (BDT = Balata dei Turchi). PALM = Palmarola (subsources Monte Tramontana and La Radica). MAR = Monte Arci (subsources: SA, SB1, SB2, SC). After Foresta Martin et al. (2020).

SAMPLES	Cl wt(%)
LIP GAB (3)	0.36
LIP CD (3)	0.34
LIP MGU (3)	0.31
PANT BDT (5)	0.50
PALM (6)	0.21
M. ARCI (12)	0.10

---

1   With an alkali (Na_2O+K_2O) excess over alumina (Al_2O_3). A peralkaline rock has by definition the peralkaline index PI = $[(Na_2O + K_2O) / Al_2O_3] > 1$.

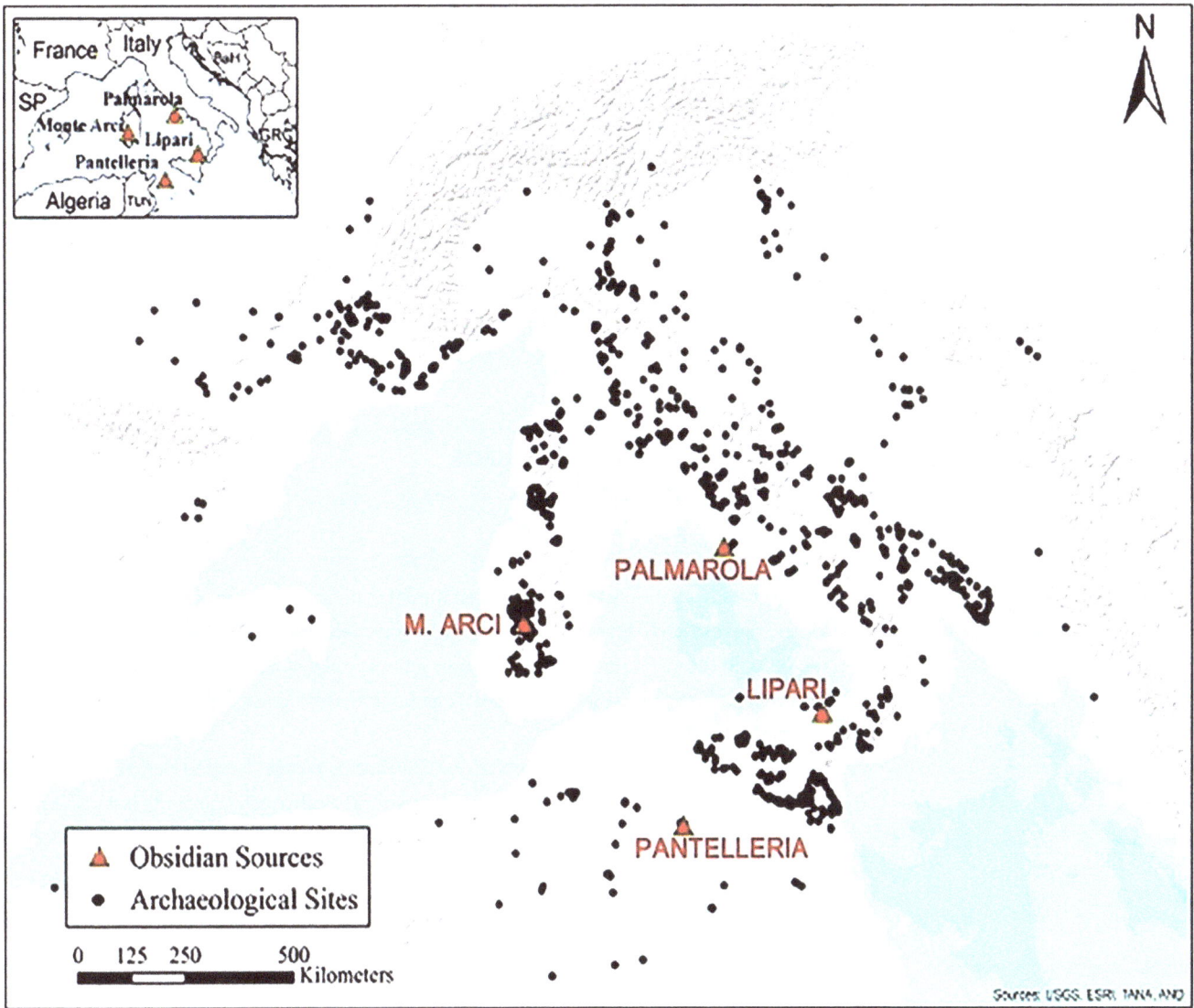

Figure 1. The Italian (Central Mediterranean) obsidian sources (red triangles) and the hundreds of prehistoric sites (black dots) where obsidian has been reported. Modified after Freund (2018).

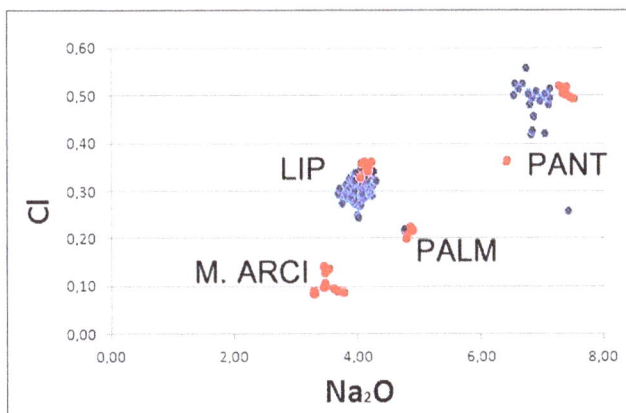

Figure 2. Clusters in the scatter plot of Cl vs. Na_2O unambiguously discriminate the four Italian obsidian sources, and also some subsources. In the plot are inserted 31 geological samples (red dots) representative of Monte Arci (M. ARCI), Palmarola (PALM), Lipari (LIP), and Pantelleria (PANT); and 174 archaeological samples (blue dots) from the island of Ustica, Sicily, attributed to Lipari (152), Pantelleria (21) and Palmarola (1). From Foresta Martin et al. (2020).

Consequential developments

Now, in this consequential work, we have investigated both the geochemical ratios Cl/Na_2O (wt%) and Cl/Na (ppm) and found that they decrease according to the sequence Lipari, Pantelleria, Palmarola, and Monte Arci, corresponding to the increasing formation ages of the obsidian outcrops. We have therefore looked for an empirical formula relating this ratio to the time of obsidian formation, with the purpose to use this geochemical parameter as a simple method to estimate the obsidian formation age. Under the methodological profile, we put forward the hypothesis, yet to be verified, that the decrease of the Cl/Na_2O or Cl/Na ratios with time depends on weathering and devitrification phenomena.

Materials and methods

Sampling

To investigate the correlation between the chlorine/sodium ratio of the obsidian samples and the respective formation ages, we have considered two groups of geological samples, based on the different geochemical techniques applied in their analyses.

The first group (Set I) includes 32 geological obsidian samples collected by F. Foresta Martin from

- Lipari, 9 samples: 3 from Vallone Gabellotto, 3 from Canneto Dentro, and 3 from Monte Guardia (there is no evidence of the prehistoric use of the last one, and it was included for its geochemical significance);
- Pantelleria, 5 samples from Balata dei Turchi;
- Palmarola, 6 samples: 3 from Monte Tramontana, and 3 from La Radica;
- Monte Arci, 12 samples: 3 for each of the sub-sources usually indicated in the literature with the abbreviations SA, SB1, SB2, SC (Tykot 1992, 2002).

The second group (Set II) includes 61 geological samples collected by R. Tykot from
- Lipari, 7 samples: 3 from Rocche Rosse and 4 from Forgia Vecchia;
- Pantelleria, 12 samples: 1 from Upper Balata dei Turchi (BdT1), 4 from Lower+Upper Balata dei Turchi (BdT2 and BdT3), and 7 from Lago di Venere;
- Palmarola, 4 samples: 2 from Monte Tramontana and 2 from Punta Vardella;
- Monte Arci, 38 samples, from the subsources SA (12), SB1 (4), SB2 (5), SC (17).

Geochemical analyses

The first group of obsidian samples (Set I) was subjected to microchemical analyses through an Electron Probe Micro Analysis (EPMA) JEOL-JXA8200, combined with Energy Dispersive Spectrometer-Wavelength Dispersive Spectrometer (EDS-WDS) and equipped with five WD spectrometers at the Istituto Nazionale di Geofisica e Vulcanologia (INGV) Laboratory in Rome. Data were collected using 15 kV accelerating voltage and 8 nA beam current.

A fragment of about 5 mm in size was detached from each sample to be analyzed. Groups of half a dozen fragments were embedded in epoxy resin stubs, abraded, and polished. The resulting mounts were ultrasonically washed in bi-distilled water and then carbon-coated before performing the microprobe analyses.

The following major and minor elements were determined: SiO_2, TiO_2, Al_2O_3, FeO_{tot}, MnO, MgO, CaO, Na_2O, K_2O, P_2O_5, Cl, F. The standards adopted for the various chemical elements are albite (Si, Al, and Na), forsterite (Mg), augite (Fe), rutile (Ti), orthoclase (K), apatite (F, P, and Ca), sodalite (Cl), celestine (S), and rhodonite (Mn). Sodium and potassium were analyzed first to further prevent alkali migration. Analytical uncertainties relative to the reported concentrations indicate that the precision was better than 5% for all cations.

The second group of geological obsidians (Set II) was subjected to Instrumental Neutron Activation Analysis (INAA). The obsidian samples were crushed, and two subsamples were prepared for INAA short and long irradiations. The short-lived elements Ba, Cl, Dy, K, Mn, and Na were measured in most samples. After decaying for ~8 days, the long irradiation samples were counted for 2,000 seconds each to measure the medium-lived elements Ba, La, Lu, Nd, Sm, U, and Yb. After three weeks, the long irradiation samples were counted for 10,000 seconds to measure long-lived elements Ce, Co, Cs, Eu, Fe, Hf, Rb, Sb, Sc, Sr, Ta, Tb, Th, Zn, and Zr. Standards including SRM-278 Obsidian Rock and SRM-1633a were similarly prepared and irradiated for calibration and quality control of the analytical data (Glascock and Ferguson 2012).

Radiometric age of obsidian samples

No radiometric analyses were performed on obsidian samples from Set I as part of this study. The radiometric dating associated with these samples was taken from the literature, in particular from the authors reported in Table 2.

On the other hand, direct radiometric analyses were carried out on obsidian samples of Set II through the $^{40}Ar/^{39}Ar$ technique. Fresh, relatively unbroken crystals were handpicked, ultrasonically cleaned, packaged in Al foil, and encapsulated in Al disks. The samples were irradiated at the McMaster University reactor, with the evolved gas purified and analyzed on the Massachusetts Institute of Technology (MIT) MAP 215-50 mass spectrometer with an electron multiplier. Measurements of the five isotopes of argon (^{40}Ar, ^{39}Ar, ^{38}Ar, ^{37}Ar, and ^{36}Ar) were corrected for system blanks, mass fractionation, and neutron-induced interferences (Flowers et al. 2006; Renne et al. 2009).

Table 2. Geological characteristics and radiometric ages of the four Central Mediterranean (Italian) obsidian sources exploited during prehistory.* The age of Canneto Dentro is unknown, but it is younger than Monte Guardia and older than Gabellotto (Lucchi et al. 2013). In the references column, the authors that have performed radiometric dating are indicated in bold.

LOCALITY	SUB SOURCES	GEOL SETTING	GEOL AGE	REFERENCES*
Monte Arci	SA SB1 SB2 SC	Within-Plate	3.6–3.2 Ma	Tykot 1992, 2002; **Montanini and Villa 1993;** **Bellot-Gurlet et al. 1999**
Palmarola	Monte Tramontana	Within-Plate; Subduction Related	1.6–1.7 Ma	**Barberi et al. 1967;** **Bellot-Gurlet et al. 1999;** Cadoux et al. 2005
Lipari	Gabellotto Canneto Dentro Monte Guardia	Subduction Related	8.7–8.4 Ka ?* 24–27 Ka	**Bigazzi and Bonadonna 1973;** **Arias et al. 1980;** **Lucchi et al. 2013;** **Zanchetta et al. 2011;** Donato et al. 2018
Pantelleria	Balata dei Turchi Salto La Vecchia Fossa Pernice	Within-Plate	127–257 Ka ?* 71–190 Ka	**Bigazzi et al., 1971;** **Radi et al. 1972;** Jordan et al. 2018; Rotolo et al. 2020

Results and discussion

Set I analyses

In a preliminary phase of this study, we focused our attention on the weight percentages of Cl and Na_2O, measured in the first obsidian group (Set I), finding that

- in each of the obsidian samples examined, the concentration of Cl is an order of magnitude lower than that of Na_2O (Table 3);

Table 3. Mean Cl and Na_2O concentrations in wt%; ratio $R = Cl/Na_2O$; and radiometric age that was taken from literature for each obsidian source. The number of analyzed (EPMA) samples are in brackets. Abbreviations as in Table 1.

SAMPLES	Cl (wt%)	Na₂O (wt%)	R = Cl/Na₂O	T = AGE (ka)
LIP GAB (3)	0.36	4.04	0.089	8,5
LIP CD (3)	0.34	4.08	0.083	(?)
LIP MGU (3)	0.31	3.71	0.084	26
PANT BDT (5)	0.50	7.28	0.069	257
PALM (6)	0.21	4.77	0.044	1600
M. ARCI (12)	0.10	3,43	0.029	3400

- regardless of the different relative proportions of Cl and Na_2O measured in the various samples, the percentage by weight of Cl with respect to Na_2O tends to decrease with the increasing age of the relative obsidian outcrop (Figure 3);

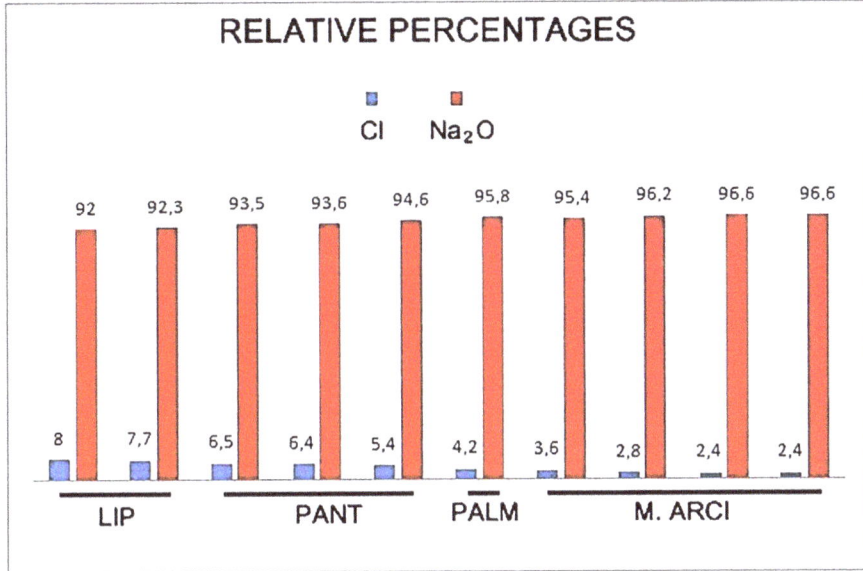

Figure 3. Mean relative percentages of Cl and Na_2O – relating to the selected obsidian outcrops of Lipari Pantelleria, Palmarola, and Monte Arci (Table 3). The proportion of Cl to Na_2O tends to decrease with the age of the outcrops.

- there is an evident negative correlation between the ratio $R = Cl / Na_2O$ and the time (T) of obsidian outcrop formation (Figure 4).

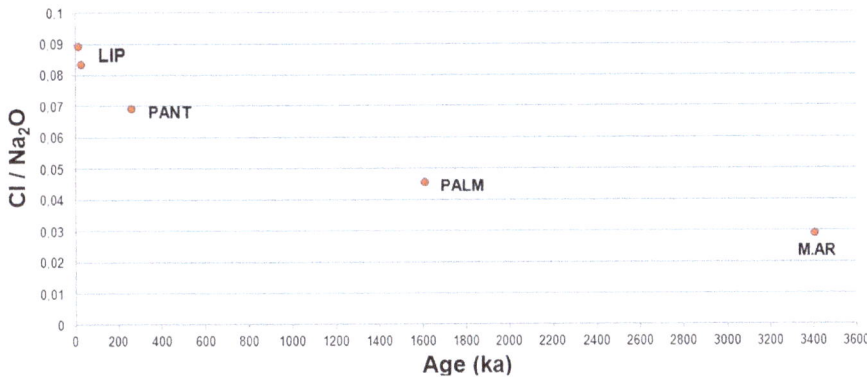

Figure 4. The mean values of $R = Cl / Na_2O$ ratio – relating to the selected obsidian outcrops of Table 3 – appear negatively correlated with the radiometric ages. A decreasing trend is evident, indicating a higher depletion rate of Cl.

We then set out to verify to what extent the ratio R can become an estimator of T. We associated each geochemical ratio $R = Cl / Na_2O$ (wt%) with the formation age of the respective obsidian deposits. The data relating to the age have been taken from the abundant literature on the subject, selecting those for which there is a concordance of values, albeit with some approximation (Table 2).

From a methodological point of view, associating our R values with radiometric ages T taken from literature could lead to errors in the definition of the best-fit curve that correlates the two parameters R and T. As a matter of fact, we do not know to which

183

part of the obsidian outcrop the radiometric data taken from the literature are related. Although they belong to a specific source, they could be part of a different deposit and have different geochemical fingerprints with respect to our samples.

Set II analyses

Awareness of the aforementioned limits and uncertainties led us to look for another set of obsidian samples on which both geochemical analyses of Cl and Na concentrations and radiometric ages were carried out (by the coauthor of this work, R. Tykot). The Cl, Na, R = Cl/Na as well as T data relating to this second group (Set II) of obsidian samples are shown in Table 4.

Table 4. Set II obsidian. Mean Cl and Na concentrations in ppm (INAA analyses, the number of analyzed samples are in brackets); ratio R = Cl/Na; and radiometric age ($^{39}Ar/^{40}Ar$ dating), for each obsidian source. Forgia Vecchia and Rocche Rosse are recent historically documented eruptions (Pistolesi et al. 2021).

Island&location	Sample	Cl ppm	Na ppm	Cl/Na	T = AGE (ka)	Error
Lipari, Forgia Vecchia*	AVE (4)	2421	29457	0.082	1.5	-
Lipari, Rocche Rosse*	AVE (3)	2189	29717	0.074	1.5	-
Pantelleria, Lago di Venere	AVE (7)	1955	37135	0.053	197	±1.8 (±0.93%)
Pantelleria, Balata dei Turchi	AVE (4)	3143	45450	0.069	236	±2.15 (±0.91)
Pantelleria, Balata dei Turchi	USF5127 (1)	3131	47678	0.066	157	±16.9 (±10.76)
Palmarola, P.ta Vardella	USF4080 (1)	1444	34724	0.042	1550	±9.4 (±0.60)
Palmarola, M. Tramontana	USF4235 (1)	1525	35283	0.043	1582	±8.9 (±0.57%)
Palmarola, P.ta Vardella	USF4092 (1)	1386	34746	0.040	1583	±9.8 (±0.62%)
Palmarola, M. Tramontana	USF4251 (1)	1284	34075	0.038	1585	±9.1 (±0.58%)
M Arci, SA	AVE (12)	852	24355	0.035	3437	±20.6 (±0.60%)
M Arci, SB1	AVE (13)	735	25021	0.030	3474	±121.5 (±3.39%)
M Arci, SB2	AVE (5)	691	23840	0.029	3332	±18.1 (±0.54%)
M Arci, SC1	AVE (9)	546	23646	0.023	3507.0	± 21.6 (± 0.62%)
M Arci, SC2	AVE (9)	533	24878	0.021	3507	±21.6 (±0.62%)

Although the Cl and Na contents were obtained with different analytical procedures and expressed in ppm, it is evident that the proportions of Cl to Na decrease passing from the youngest obsidian deposits to the oldest ones, with some exceptions found in several sub-sources. Thus, having available two sets of samples analyzed with different methods, we searched the mathematical law that expresses the correlation between R and T in the two cases and compared the two results.

The search for a mathematical law

We then tried to verify if the behavior of the ratio R with the age T can be well represented by a mathematical law, which can also provide information about physical modeling for the change of R. Our starting assumption was that the typical time scale of the R evolution depends upon a power of the ratio itself:

$$(1/R)(dR/dt) = -B R^n \qquad (1)$$

where B is a positive constant and n is a non-negative number. This equation can be resolved by means of a simple integration for the separation of variables:

$$\int dR/R^{n+1} = -\int B \, dt$$

that, for $n = 0$, corresponds to an exponential decay

$$R(t) = A \, exp(-Bt) \qquad (2)$$

where the integration constant A is the initial value of ratio R_0, while for n different from 0, one obtains after a simple algebraic manipulation

$$R(t) = 1/(A + nBt)^n \qquad (3)$$

with $A = 1/(R_0)^n$, that in geophysics is known as Omori-Utsu law, a generalization of the previous Omori law (Omori 1894), which is obtained in the particular case of $n = 1$,

$$R(t) = 1/(A + Bt) \qquad (4)$$

and $A = 1/R_0$.

We computed several numerical best fitting of these formulae to the data by a least mean square linear minimization and found that Omori's law of Eq. (4) is the most successful in describing the entire data set. This law was introduced on an empirical basis for describing the frequency of aftershocks following a strong earthquake (Omori 1894; Utsu 1961; Utsu and Ogata 1995). Its physical interpretation is still not fully understood, although it is a general view that it is related to some rupture mechanism in the rocks of the Earth's crust. As it will be discussed in Section 3.6, it is possible to devise a likely physical interpretation for its use in the present context. Eq. (4) can be easily inverted using R to obtain an estimate of the age of obsidian samples:

$$t = (1/B)(1/R - A) \qquad (5).$$

It is interesting to note that for high ages of the obsidian—which correspond to low values of R, such that $1/R$ is much higher than A—the age estimate tends to be independent on the parameter A, which is related to the initial value of R. We can define a typical time scale of R evolution that is:

$$\tau = A/B \qquad (6)$$

that is the time necessary to reach a value of $R(\tau) = R_0/2$, the half of the initial one.

The best fit curve

Here, we present the results of the best fitting of Eq. (4) to the two different sets of data. The first set is the one given in Table 1, indicated as Set I, which is the same already presented at the 2021 International Obsidian Conference and described in paragraph 3.1. The best fit curve (in red color) is shown in Figure 5, where the age of samples is reported in units of 1,000 years. The best-fit values of the parameters are also given in the figure. In the same figure, we also reported the exponential best curve (dashed green line): it is clear from the comparison of the two curves that the Omori law is closer to the data than the exponential law, which has an rms of the residuals that is higher than the other one by a factor of about two.

Figure 5. Two best-fit curves to the Set I data (blue-filled circles): the solid red line is the Omori law, while the green dashed line is the exponential law. Numerical formulae with best-fit values of the parameters are written inside the figure.

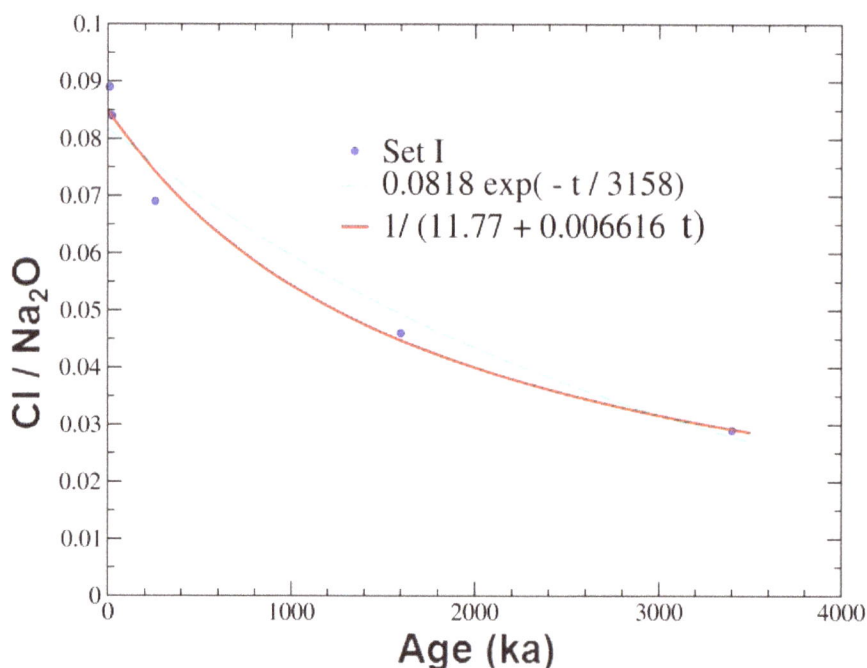

The same law also gives the best fit to the Set II data as shown by the turquoise curve in the Figure 6 plot. Interestingly, parameter *B* has nearly equal best fit values for the two fits, while parameter *A* is about 13% higher. This implies that for high ages the corresponding values are not largely different, while for small ages the samples of Set II have lower values. For comparison, we reported in the same Figure 6 the best fit curve (dashed red line) of the Set I data that makes evident this difference.

Taking into account that the differences between the two data sets may be affected by systematic uncertainties likely due to the techniques used, we considered an additional Set C obtained by joining together the two previous ones. The best fit of the Omori law applied to this "combined" set is shown in Figure 7: the resulting curve is clearly intermediate between the two other curves and can represent the most acceptable solution

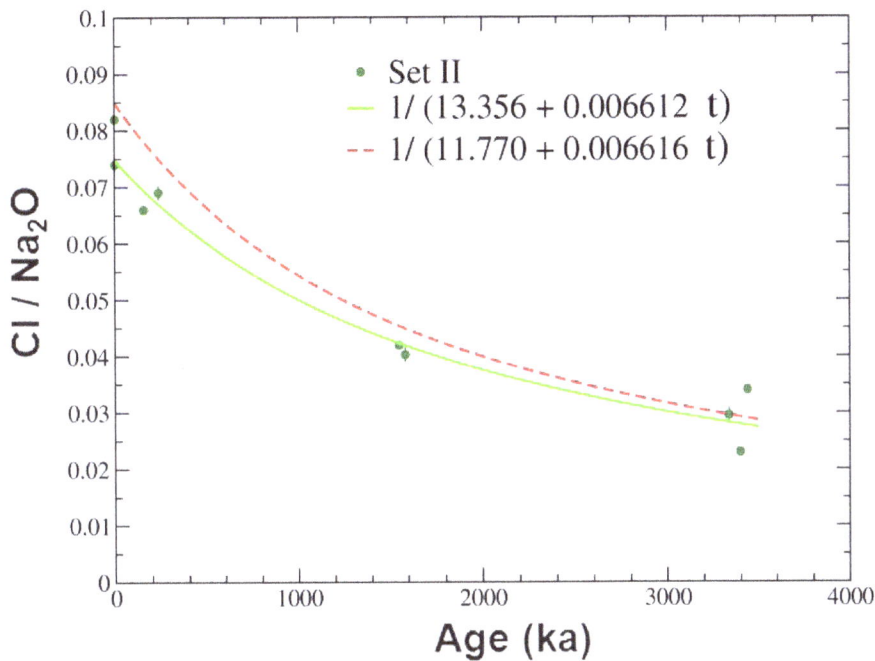

Figure 6. The best fit of the Omori law to the Set II data (green-filled circles): the green solid line is the best fit; the Set I is also shown for comparison. Numerical formulas with best-fit values of the parameters are written inside the figure.

for the time evolution of the ratio R. Using the parameter values reported in Figure 7, we can obtain from Eq. (5) the following practical formula for estimating the age of a sample from its value of R:

$$t = 149.3 (1/R - 12.68) \tag{7}$$

in units of ka. Of course, for R values > 0.0789, this formula cannot be applied because it would give a negative age. Such a result must be interpreted in terms of high uncertainty in the initial composition of the samples, and therefore for values higher than 0.078 one can only establish a higher limit to the age of about 22 ka.

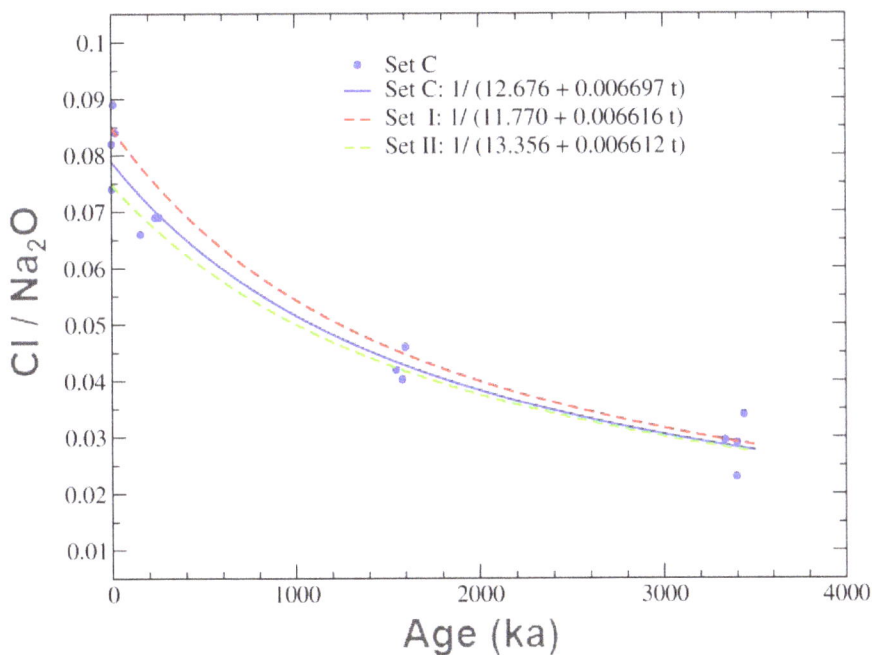

Figure 7. The best fit of the Omori law to the Set C data (blue-filled circles): the blue solid line is the best fit; those of Set I and II are also shown for comparison. Numerical formulas with best-fit values of the parameters are written inside the figure.

We also performed the same analysis on a set in which the data with the lowest and highest values were omitted from the Lipari and Monte Arci samples, and the resulting parameters were practically unchanged. As a further step, we omitted all the Lipari data, whose R values might be highly depending on the initial condition because of their young age; taking into account only the Pantelleria, Palmarola, and Monte Arci samples, we obtained the best fit curve very close to that of Set II.

In conclusion, Eq. (7) currently appears to be the most reliable relation between the Cl/Na ratio of the studied obsidians and their ages.

Testing the method on non-Mediterranean obsidian

As shown in the previous sections, the correlation found for Italian obsidians between their Cl/Na_2O ratio and time of formation can allow us to use this parameter to estimate the age of obsidian sources. However, the observed decrease of Cl/Na_2O with time could also be only apparent, and the different ratios shown by the Italian obsidian sources could be due to other factors such as different magma sources and/or different evolutionary processes.

Due to the limited number of sources in the Mediterranean area for which both chemical analyses of Cl and age are available, in order to have a first test of the possible use of R as an estimate of the obsidian age, we have applied Eq. 7 to the obsidians of Sierra de las Navajas (State of Hidalgo, Mexico). EPMA analyses of these obsidians are reported in the recent paper of Donato and colleagues (2022), and their average Cl and Na_2O contents are summarized in Table 5.

Table 5. Average Na_2O and Cl concentrations of Sierra de las Navajas obsidians; obtained from data in Donato et al. (2022).

LOCATION	Na₂O (wt%)	Cl (wt%)	R = Cl/Na₂O	T = AGE (ka)
Sierra de Las Navajas	4.9394	0.2023	0.0410	1752.6743

The average R-value for these obsidians is about 0.041. Applying Eq. 7, it gives an age of formation of 1.75 Ma. The obsidians of Sierra de las Navajas come from a deposit formed during the early stages of the volcano building. The volcanic edifice was successively involved in a sector collapse, producing a debris-avalanche deposit (Lighthart Ponomarenko 2004). The age of the Sierra de las Navajas volcano is not known; however, according to Nelson and Lighthart (1997), the products of the debris avalanche directly overlie basaltic lavas whose K/Ar age is 1.8 +- 0.4 Ma. An age younger than 2 Ma is also suggested by Lighthart Ponomarenko (2004) for the Sierra de las Navajas volcanic edifice. These ages are in agreement with the formation age obtained by applying Eq. 7. This represents, therefore, a first positive test for the possible use of the Cl/Na_2O ratio to roughly infer obsidian ages.

A possible explanation

In an attempt to give a possible explanation of the physical and geochemical processes that lead to the decrease of R with the age of obsidian outcrops, we must keep in mind some degradation processes to which volcanic glasses, in general, are subject. Obsidian is a silicate (SiO_2 > 65 wt%) super-cooled liquid characterized by a poorly ordered internal structure, with roughly linked SiO_4 tetrahedra that form a polymerized network and appreciable intermolecular space. On a geological time scale, obsidians are thermodynamically unstable and prone to glass network destruction mainly due to environmental-alteration phenomena, i.e. weathering (Fisher and Schmincke 1984 and references therein). Alteration begins with the rapid absorption of thin H_2O layers on the glass surface, followed by slow diffusion of the water inside the glass, favored by the opening of mini-cracks and the deconstruction of the polymerized network (Friedman and Long 1976; Friedman and Trembour 1978; Jezek and Noble 1978). This so-called hydration process, which is an exponential function of time and temperature, in obsidians goes on up to an H_2O saturation limit of about 3%; and it underlies the dating methods of flaked obsidian tools where it is possible to identify and measure the hydration layer (Friedman and Smith 1960; Friedman and Long 1976; Friedman and Trembour 1978).

During the water diffusion phase inside the glass, the most relevant chemical changes consist of an Na decrease and K increase by ion exchange with the groundwater, while volatile components such as Cl tend to be lost (Jezek and Noble 1978). The study of the alterations on volcanic ashes has ascertained that the micro fracturing of the glass in the absence of water is very slow, while the circulation of hot water and high environmental temperatures accelerate the hydration and the network dissolution processes (Zielinsky 1980). Volcanic glasses with higher concentrations of Si and Al network-forming elements exhibit greater resistance to these alteration processes, as shown by comparative experimental studies (Fisher and Schmincke 1984 and references therein).

Devitrification of obsidian glass can also contribute to a further loss of both Cl and Na in very old obsidian, as that process preferentially reorganizes silica like cristobalite and potassium as K-feldspar (Bullock et al. 2017) in white spherulites and blade-like crystals (Gimeno 2003). Devitrification increases interstitial voids in the obsidian, enhancing water permeation and mobile elements loss, which have not been accommodated in crystal reticula (Gimeno 2003).

Putting it all together, Cl and Na appear to be linked by a common fate within obsidians. The solubility of Cl tends to increase in alkaline and peralkaline melts, where Cl pairs preferentially with Na (Carroll and Webster 2018) and the concentrations of both elements are positively correlated. But, just as both of these elements enter and pair in the rhyolitic melts, they both exit due to the hydration/alteration processes to which volcanic glasses are subject. However, chlorine diffusivity is probably higher than that of sodium and can occur also in the absence of water. Therefore, also in "fresh," not hydrated obsidians not experiencing a significant Na loss, a remarkable chlorine loss can occur with time. Regardless of the different genesis and the differences in chemical compositions that characterize the obsidian deposits considered in this study, it seems that the different loss rates of Cl and Na can be the explanation for the inverse correlation between R and T—and make it possible to use R as an empirical estimator of the age of an obsidian outcrop.

It is also a possible explanation for the success of the Omori law in describing the time evolution of the Cl to Na ratio. To interpret our results, we recall that a simple phenomenological interpretation of the Omori law was given by Guglielmi (2016)—in analogy with the recombination process of oppositely charged particles in the ionospheric gas—who proposed that an earthquake occurs in an "active" fault at which opposite shear stresses are present. Other researchers, for instance Mallick and colleagues (2009), have shown that microfractures occurring at a constant deformation in an indention experiment on a glass cylinder are also finely described by an Omori law, which can be related to a thermally activated rupture process. More recently, Roy and Hatano (2018) studied the creep-like behavior of fractures in heterogeneous media under a constant load and demonstrated that they are described by the Omori-Utsu law. In the present context we can, therefore, assume the simple first-order approximation that the decreasing rate of the ratio R is proportional to the frequency of the microfractures in the obsidian and that this frequency is described by the Omori law.

Conclusion

It has been observed that, in the Italian obsidian, values of the ratio $R = Cl/Na_2O$ (wt%), and Cl/Na (ppm) of old obsidian sources (e.g., Monte Arci, 3.4 Ma, $R = 0.029$) are much lower than the young ones (e.g., Lipari, 22–10 ka, $R = 0.084$–0.089). Different concentrations in the various elements can be related to the different geodynamic settings (and, consequently, geochemically different magma sources) in which the obsidian has been emplaced. However, such a systematic decrease of an element ratio with time could also be due to a preferential loss of the more volatile component (Cl) with respect to the more stable Na, due to micro fracturing of the obsidian with time. If this is true, the Cl/Na_2O ratio could be used as an empirical age estimator of the obsidian source, capable of being also a discriminating tool for obsidian, which, although belonging to different magma sources, nonetheless have similar elemental compositions (Shackley et al. 2017).

For the Italian sources, the decreasing evolution of R is well described by means of the inverse law in Eq. (4) and a best-fit evaluation of the parameters results in the formula in Eq. (7). The accuracy of this method depends on the reliability of calibrators—more precisely on the scatter of R in samples from the same source as in the case of Monte Arci obsidian.

It is necessary to test the method on a higher number of obsidian sources whose age and chemical composition are known to understand if its validity is limited to the Italian obsidian or if it can be used universally. We attempted a first independent verification of our Eq. (7) to infer the age of different obsidian by applying it to samples from the Sierra de las Navajas (Mexico) obsidian source, for which an average value of $R = 0.041$ has been found. The age of ca. 1.7 Ma is in agreement with an age younger than 1.8–2 Ma, generally accepted for this source.

In this work, we have not tested the method on archaeological samples. However, our previous studies demonstrated that Cl and Na contents of archaeological samples are perfectly coincident with those of the geological samples and can be therefore used as discriminating elements between Mediterranean sources (Foresta Martin et al. 2020). A possible limit to the application of this method is the necessity to analyze fresh surfaces, which, in the case of archaeological samples, means that they cannot be perfectly preserved.

The Cl and Na_2O contents of obsidians can be obtained by several analytical methods, including the quick and poorly destructive or non-destructive Energy Dispersion System (EDS) or Wavelength Dispersion System (WDS) associated with an electron microscope or microprobe. If the relationship between age and Cl/Na_2O ratio is confirmed, therefore, the proposed method can be used to get the first estimate of the age of the obsidians through the use of a very simple and quick analysis.

Acknowledgments

We express our gratitude to Professor Pasquale Mario Nuccio for the fruitful discussions concerning the possible geochemical interpretations of the law described in this chapter; and to the organizers of the International Obsidian Conference (IOC) held in Berkeley, California, April 30–May 2, 2021, who offered us a suitable audience to collect further analytical data.

References

Arias, C., G. Bigazzi, and F.P. Bonadonna
1980   Studio cronologico e paleomagnetico di alcune serie sedimentarie dell'Italia appenninica. *Contributi alla realizzazione della Carta Neotettonica d'Italia.* Pub. no. 356 of the Geodynamic Finalized Program – Neotectonic Sub-Project: 1441–1448.

Barberi, F., S. Borsi, G. Ferrara, and F. Innocenti
1967   Contributo alla conoscenza vulcanologica e magmatologica delle isole dell'Arcipelago Pontino. *Memorie della Societa Geologica Italiana* 6: 581–606.

Bellot-Gurlet, L., Bigazzi, G., Dorighel, O., Oddone, M., Poupeau, G., and Yegingil, Z.
1999   The fission-track analysis: An alternative technique for provenance studies of prehistoric obsidian artefacts. *Radiation Measurements* 31(1–6): 639–644.

Bigazzi, G., F. P. Bonadonna, G. Belluomini, and L. Malpieri.
1971   Studi sulle ossidiane italiane. IV. Datazione con il metodo delle tracce di fissione. *Bollettino della Società Geologica Italiana* 6: 581–606.

Bigazzi, G., and F. P. Bonadonna
1973   Fission track dating of the obsidian of Lipari Island (Italy). *Nature* 242(5396): 322–323.

Bonifacie, M., N. Jendrzejewski, P. Agrinier, E. Humler, M. Coleman, and M. Javoy.
2008   The chlorine isotope composition of Earth's mantle. *Science* 319(5869): 1518–1520.

Bullock, L. A., R. Gertisser, and B. O'Driscoll
2017   Spherulite formation in obsidian lavas in the Aeolian Islands, Italy. *Periodico di Mineralogia* 86: 37–54.

Cadoux, A., D. L.Pinti, C. Aznar, S. Chiesa, and P. Y. Gillot.
2005   New chronological and geochemical constraints on the genesis and geological evolution of Ponza and Palmarola volcanic islands (Tyrrhenian Sea, Italy). *Lithos* 81(1–4): 121–151.

Carroll, Michael, and James D. Webster
2018   Solubilities of sulfur, noble gases, nitrogen, chlorine, and fluorine in magmas. In *Volatiles in magmas*, edited by Michael R. Carrol, and John R. Holloway, pp. 231–280. De Gruyter, Boston.

Donato, P., L. Barba, R. De Rosa, G. Niceforo, A. Pastrana, S. Donato, G. Lanzafame, L. Mancini, and G. M. Crisci.

  2018   Green, grey and black: A comparative study of Sierra de las Navajas (Mexico) and Lipari (Italy) obsidians. *Quaternary International* 467: 369–390.

Donato, P., S. Donato, L. Barba, G. M. Crisci, M. C. Crocco, M. Davoli, R. Filosa, V. Formoso, G. Niceforo, A. Pastrana, A. Solano, and R. De Rosa.

  2022   Influence of Chemical Composition and Microvesiculation on the Chromatic Features of the Obsidian of Sierra de las Navajas (Hidalgo, Mexico). *Minerals* 12(2): 177.

Fisher, R. V., and H. U. Schmincke

  1984   *Pyroclastic Rocks.* Springer, Berlin, Heidelberg.

Flowers, R. M., K. H. Mahan, S. A. Bowring, M. L. Williams, M. S. Pringle, and K. V. Hodges

  2006   Multistage exhumation and juxtaposition of lower continental crust in the western Canadian Shield: Linking high-resolution U-Pb and ^{40}Ar/^{39}Ar thermochronology with pressure-temperature-deformation paths. *Tectonics* 25(4): 1–20.

Foresta Martin, F., S. G. Rotolo, M. Nazzari, and M. L. Carapezza.

  2020   Chlorine as a discriminant element to establish the provenance of Central Mediterranean Obsidians. *Open Archaeology* 6(1): 454–476.

Freund, K. P.

  2018   A long-term perspective on the exploitation of Lipari obsidian in Central Mediterranean prehistory. *Quaternary International* 468a: 109–120.

Friedman, I., and W. Long.

  1976   Hydration Rate of Obsidian: New experimental techniques allow more precise dating of archeological and geological sites containing obsidian. *Science* 191(4225): 347–352.

Friedman, I., and F. Trembour.

  1983   Obsidian hydration dating update. *American Antiquity* 48(3): 544–547.

Gimeno, D.

  2003   Devitrification of natural rhyolitic obsidian glasses: petrographic and microstructural study (SEM+ EDS) of recent (Lipari Island) and ancient (Sarrabus, SE Sardinia) samples. *Journal of Non-crystalline Solids* 323(1–3): 84–90.

Glascock, M. D., and Ferguson, J. R.
2012   Report on the Analysis of Obsidian Source Samples by Multiple Analytical Methods. Archaeometry Lab, University of Missouri Research Reactor, Columbia, Missouri.

Guglielmi, A. V.
2016   Interpretation of the Omori law. *Izvestiya, Physics of the Solid Earth* 52(5): 785–786.

Jezek, P. A., and D. C. Noble.
1978)   Natural hydration and ion exchange of obsidian; an electron microprobe study. *American Mineralogist* 63(3–4): 266–273.

Jordan, N. J., S. G. Rotolo, R. Williams, F. Speranza, W. C. McIntosh, M. J. Branney, S. and Scaillet.
2018   Explosive eruptive history of Pantelleria, Italy: Repeated caldera collapse and ignimbrite emplacement at a peralkaline volcano. *Journal of Volcanology and Geothermal Research* 349: 47–73.

Lanzo, G., P. Landi, and S. G. Rotolo.
2013   Volatiles in pantellerite magmas: A case study of the Green Tuff Plinian eruption (Island of Pantelleria, Italy). *Journal of Volcanology and Geothermal Research* 262: 153–163.

Le Bourdonnec, F. X., G. Poupeau, and C. Lugliè.
2006   SEM-EDS analysis of western Mediterranean obsidians: A new tool for Neolithic provenance studies. *Comptes Rendus Geoscience* 338(16): 1150–1157.

Le Bourdonnec, F. X., J. M. Bontempi, N. Marini, S. Mazet, P. F. Neuville, G. Poupeau, and J. Sicurani.
2010   SEM-EDS characterization of western Mediterranean obsidians and the Neolithic site of A Fuata (Corsica). *Journal of Archaeological Science* 37(1): 92–106.

Lighthart Ponomarenko, A.
2004   The Pachuca obsidian source, Hidalgo, Mexico: A geoarchaeological perspective. *Geoarchaeology* 19: 71–91.

Lowenstern, J. B.
1994   Chlorine, fluid immiscibility, and degassing in peralkaline magmas from Pantelleria, Italy. *The American Mineralogist* 79: 353–369.

Lucchi, F., C. A. Tranne, F. Forni, and P. L. Rossi.
2013   Geological map of the Island of Lipari, scale 1:10000 (Aeolian archipelago). In *The Aeolian Islands Volcanoes*, edited by Lucchi, F., A. Peccerillo, J. Keller, C. A. Tranne, and P. L. Rossi, pp. 113–153. Geological Society of London.

Mallick, N., S. Ciliberto, S. G. Roux, P. Di Stephano, and L. Vanel.
2009   Aftershocks in thermally activated rupture of indented glass. In *Proceedings of the 12th International Conference on Fracture, ICF-12* (July), pp. 12–17, Ottawa, Ontario, Canada.

Montanini, A., and I. M. Villa.
1993 ^{40}Ar/^{39}Ar chronostratigraphy of Monte Arci volcanic complex (western Sardinia, Italy). *Acta Vulcanologica* 3: 229–233.

Nelson, S. A., and A. L. Lighthart.
1997   Field excursion to the Sierra Las Navajas, Hidalgo, Mexico – A Pleistocene Peralkaline rhyolite complex with a large debris avalanche deposit. In *II Convención sobre la Evolución Geologica de México y Recursos Asociados. Libro-guía de las excursiones geológica*, pp. 89–96.

Omori, F.
1894   On after-shocks of earthquakes. *Journal of the College of Science, Imperial University of Tokyo* 7: 111–200.

Pistolesi, M., M. Rosi, A. B. Malaguti, F. Lucchi, C. A. Tranne, F. Speranza, P. G. Albert, V. C. Smith, A. Di Roberto, and E. Billotta.
2021   Chrono-stratigraphy of the youngest (last 1500 years) rhyolitic eruptions of Lipari (Aeolian Islands, Southern Italy) and implications for distal tephra correlations. *Journal of Volcanology and Geothermal Research* 420: 107397.

Renne, P. R., A. L. Deino, W. E. Hames, M. T. Heizler, S. R. Hemming, K. V. Hodges, A. P. Koppers, D. F. Mark, L. E. Morgan, D. Phillips, B. S. Singer, B. D. Turrin, I. M. Villa, M. Villeneuve, and J. R. Wijbrans.
2009   Data reporting norms for ^{40}Ar/^{39}Ar geochronology. *Quaternary Geochronology* 4: 346–352.

Radi, G., G. Bigazzi, and F. Bonadonna.
1972   Le tracce di fissione. Un metodo per lo studio delle vie di commercio dell'ossidiana. *Origini* 6: 155–169.

Roy, S., and Hatano, T.
2018   Creeplike behavior in athermal threshold dynamics: Effects of disorder and stress. *Physical Review E* 97(6): 062149.

Rotolo, S. G., M. L. Carapezza, A. Correale, F. Foresta Martin, G. Hahn, A. G. E. Hodgetts, M. La Monica, M. Nazzari, P. Romano, L. Sagnotti, G. Siravo, and F. Speranza.
2020   Obsidians of Pantelleria (Strait of Sicily): A petrographic, geochemical and magnetic study of known and new geological sources. *Open Archaeology* 6(1): 434–453.

Shackley, M. S., L. Morgan, and D. Pyle.

2018   Elemental, isotopic, and geochronological variability in Mogollon-Datil volcanic province archaeological obsidian, southwestern USA: Solving issues of intersource discrimination. *Geoarchaeology* 33: 486–497.

Tykot, R. H.

1992   The sources and distribution of Sardinian obsidian. In *Sardinia in the Mediterranean: A footprint in the sea*, edited by R. H. Tykot and T. K. Andrews, pp. 57–70. Sheffield Academic Press, Sheffield.

Tykot, R. H.

2002   Chemical fingerprinting and source tracing of obsidian: The central Mediterranean trade in black gold. *Accounts of Chemical Research* 35(8): 618–627.

Utsu, T.

1961   A statistical study on the occurrence of aftershocks. *Geophysical Magazine* 30: 521–605.

Utsu, T., and Ogata, Y.

1995   The centenary of the Omori formula for a decay law of aftershock activity. *Journal of Physics of the Earth* 43(1): 1–33.

Zanchetta, G., R. Sulpizio, N. Roberts, R. Cioni, W. J. Eastwood, G. Siani, B. Caron, M. Paterne, and R. Santacroce.

2011   Tephrostratigraphy, chronology and climatic events of the Mediterranean basin during the Holocene: an overview. *The Holocene* 21(1): 33–52.

Zielinski, R. A.

1980   Stability of glass in the geologic environment: some evidence from studies of natural silicate glasses. *Nuclear Technology* 51(2): 197–200.

CHAPTER 10

Open Access Data Repositories for Obsidian Geochemistry: Prospects and Impediments

NICHOLAS TRIPCEVICH, MICHAEL D. GLASCOCK, AND ERIC C. KANSA

Nicholas Tripcevich Archaeological Research Facility, UC Berkeley. Berkeley, CA 94720, tripcevich@berkeley.edu

Michael D. Glascock University of Missouri, Research Reactor Center, Columbia, MO 65211 glascockm@missouri.edu

Eric C. Kansa Alexandria Archive Institute / Open Context, San Francisco, CA 94127 eric@opencontext.org

Abstract

This chapter proposes a novel approach to disseminating obsidian geochemistry data. It is common today to find geochemical results (complete or in summary form) distributed along with a publication much as was practiced 50 years ago in paper form. Using a web repository approach and inspired by a geospatial system employed by the US Geological Survey, among others, we propose that geochemical analyses should be hosted at permanent web locations so that these data can be built into other web-based analysis systems. Further, we suggest that version control software can be used to update geochemistry collections online and track analysis sessions.

Introduction

More than 50 years have passed since geochemical sourcing of obsidian began at a few national laboratories, and much has changed in how analyses are conducted and results reported. Reflecting these new realities of research, this chapter presents a model where analyses make greater use of open access publishing repositories and, in some cases, version control software.

Until recent years, the dominant method for circulating obsidian geochemistry results was to publish the results in regional or scientific journals together with a discussion of the implications for regional archaeology. Furthermore, when printed on paper many publishers were averse to publishing complete data sets due to space constraints, and so only summaries (mean and standard deviation) were included in many publications. While this ensures that the summary results and analysis will circulate and be read, it stands in contrast to the granular detail and cumulative nature of the data needed for obsidian geochemistry research in a region. That is, if one is expected to contextualize every dataset with an essay reflecting on the significance in the region and then find a publisher to circulate it, it limits the sharing of the data that should happen—and it results in analysts saving up data sets for the eventual "big publication" instead of circulating them digitally in the tradition of many physical sciences. Obstacles to data dissemination are numerous, including data guarding or lack of funding or available time for rendering datasets comparable with others, and these interfere with the goals of inter-lab comparisons and reproducibility in science (Stark 2018). Inspired by the open data and reproducibility movement in recent years (Kansa and Kansa 2021; Marwick 2017), we describe a repository hosting obsidian data and outline a system for incorporating new X-ray fluorescence (XRF) datasets into a version-controlled repository. The infrastructure for data sharing on the internet has much to offer geochemistry, and obsidian source datasets are a discrete group that can be used to explore the immediate benefits of this approach.

Expanding analyses of many collections with inter-instrumental comparison

With the greater availability of spectrometers and the demand for non-destructive analytical methods, a wider array of researchers has gained access to portable XRF (pXRF). While still relatively costly and not always durable, pXRF units have the advantage of being small enough to be transported on airplanes to the collections found in labs and museums, and even to field sites running on battery power. This has greatly expanded the number of analyses occurring annually and allowed for the systematic analysis of collections worldwide. While use of the reference samples and standards on a single instrument is preferable, it does not scale. As a result, inter-instrument coordination and comparability are a necessary step for expanding obsidian research going forward.

Inter-instrumental comparability has been an often-noted challenge in these studies. Geological standards such as the Obsidian Rock (SRM-278) or Rhyolite Glass Mountain (RGM) have been used as check standards to evaluate comparability between instruments and drift or error within a single machine (Glascock et al. 1998; Govindaraju 1994). To control for high and low values within each element of interest in XRF, in the past decade, two reference sets have become available for analysts (Frahm 2019; Glascock and Ferguson 2012; Johnson et al. 2021) that represent an important step forward in generating comparable inter-instrumental values. These calibration sets have been developed together with "recommended values" tables for empirical calibration (Glascock and Ferguson 2012) or using custom factors on Fundamental Parameters-based XRF instruments (Frahm 2019). This allows analysts to target specific concentration values per element for known standards or sources, and to derive valid numbers for artifacts without necessarily having the target chemical values for known obsidian sources in a region. This development highlights the need for coordinated data-sharing in terms of reference datasets and the chemistry of obsidian sources as measured by these comparable instruments.

Source assignment challenges

A primary question in the analysis of obsidian artifacts is source assignment to a specific geological obsidian flow. These assignments are typically localized (reflecting mobility and transport in antiquity in a given region). A common practice in our discipline is that archaeologists with instrumentation and a specialty in geochemistry will make the assignments by way of comparison with source samples or by way of trace-element reference data that was, under ideal circumstances, derived from source samples that they have run on their instruments.

Databases of source locations and names are available, and datasets could be effectively organized into a web-based Geographical Information System (GIS). These regionally important sources have quarrying and workshop sites nearby, and therefore are protected archaeological features by definition. There are methods for obfuscating site locations that could be employed without substantially diminishing the possibilities for regional spatial analysis. For example, the OpenContext repository uses the Quadtree Indexing approach where precision locations are aggregated into cells that are typically rendered using choropleth shading to indicate density of sites in each cell. These are shared at a pre-defined spatial resolution for the grid cells. In one example at OpenContext at the Zoom level 11, the spatial information can be shared at the relatively coarse scale of 17 km cells using the Mercator projection; but for features deemed less sensitive, location could be shared on a finer scale, with perhaps 3 km cells. In sum, the regional spatial relationships are maintained, and viewers of the web maps gain an understanding of the density of cultural features without revealing precision site locations. Given the expanding number of instruments and demand for non-destructive analysis, we present here two approaches for improving access to reference data.

Reference data from Neutron Activation Analysis

The results of Neutron Activation Analysis (NAA) are stable and unlikely to change, so they are appropriate for sharing in a permanent repository. Data sets have been made available on laboratory websites such as that of the University of Missouri Research Reactor (MURR) [https://archaeometry.missouri.edu/murr_database.html], as well as in permanent online repositories, such as OpenContext and the Digital Archaeological Record (tDAR). It is worth noting that results of many XRF studies are also available on repositories, such as the 500+ reports distributed on the University of California eScholarship system by M. Steven Shackley (https://escholarship.org/uc/arf_xrfreports).

The Andean Obsidian Geochemistry Project (https://doi.org/10.6078/M7VH5M0N) on the OpenContext repository hosts over 1,300 records of NAA, which runs from the Archaeometry lab at MURR. These are available in user delimited regions as data downloads or through the web software without pre-registration. This approach facilitates integration into web-based visualization software such as SourceXplorer (McMillan 2021) or other R or Python-based web visualization tools like Jupyter notebooks or built into a Geographical Information System. A well-developed example of just such a geospatial system is provided by the US Geological Survey in the National Geochemical Database (https://mrdata.usgs.gov/ngdb/rock/), in which the geochemistry for over 400,000 igneous rock samples (mostly via emission spectrography) are provided at a stable, web geospatial archive. This allows researchers to compare their field samples to the existing geochemistry from samples acquired nearby.

While the use of chemical concentrations derived from NAA in assigning obsidian artifacts to sources based on XRF data is more complex (Glascock 2011, 2020), a number of elements are comparable (in particular, Ba, Rb, Zr, and Sr > 10 ppm), and these can serve as the basis for assigning samples to sources with an XRF instrument, with a calibration or results adjusted to match concentrations derived from other geochemical methods.

OpenContext interface

OpenContext is an open-data publishing service for archaeology. The OpenContext API provides access to a variety of structured queries without requiring a sign-in, and the web interface of OpenContext allows analysts to subset the NAA data spatially using the map interface (Figure 1) and include source geochemistry within the research area of interest. In the example shown here, square cells overlain on the national boundaries (OpenStreetMap) show density of source samples by cell color, and the selection box bounds a subset of the full collection. A subsequent screen provides the coordinates of the selection, and, via the Download button, the chemistry of 98 source samples in this example are provided. These can be acquired either as a CSV (table) or a GeoJSON (GIS data) for inclusion in geospatial or statistical software and for comparison with the chemistry of obsidian artifacts from the region.

Open and reproducible science

These approaches contribute to a development in scientific inquiry where data and code are accessible and made available by developers for researchers and consumers. While instrumentation and development require significant capital investment, the entire process—the instrument construction and collimation, the design and thickness of filters, the algorithms that interpret the spectra, and the calibrations used to convert the spectra to ppm concentrations—would be published and accessible to researchers.

Unfortunately, these instruments are designed for a narrow industrial market, so many of the assumptions built into calibrations and fundamental parameters algorithms

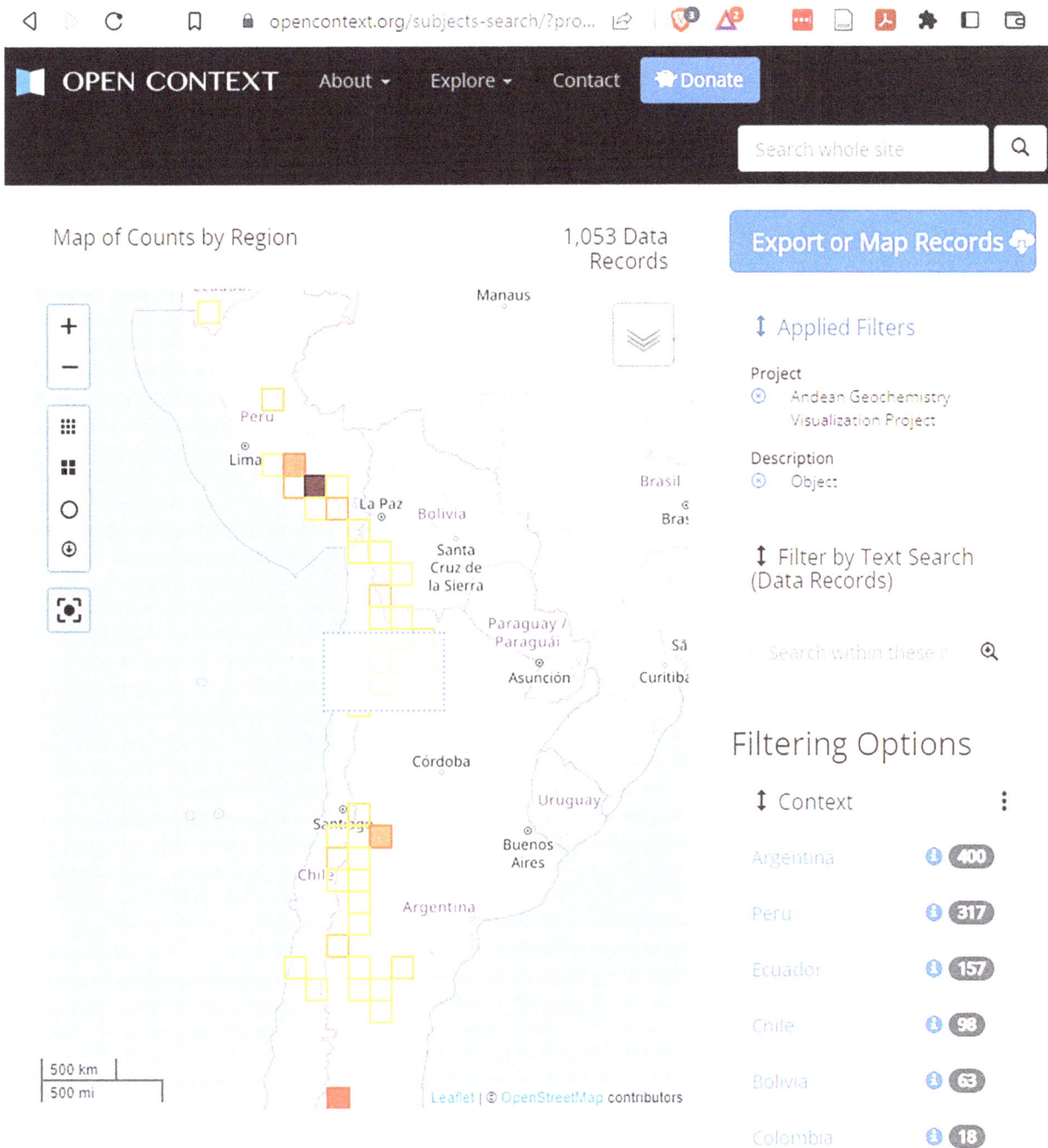

Figure 1. The project on OpenContext showing the density of source sample data by cell color and an Area-of-Interest selection box for exporting a subset of the full collection.

are not published and are approached as trade secrets for this market. Despite these gateways in instrumentation, the larger field of obsidian studies can continue to use principals from reproducible and open science to share data and define the geochemical regions and the complex areas where overlapping geochemistry requires special attention.

XRF geochemistry research with a commitment to reproducible science contributes to current approaches in data science (Brunsdon and Comber 2021) where transparent methods and results are available for a variety of analytical approaches and unlink the results of the research from commercial hardware and software constraints imposed by existing workflows.

Version control for X-ray fluorescence values from particular obsidian sources

In order to assign artifacts to sources in an unfamiliar region, one must typically reach out to archaeologists who have published on obsidian in that region, and, with any luck, one will receive a table of analysis results for sources in the region. While some have been published or hosted in online repositories such as those by Shackley (https://escholarship.org/uc/arf_xrfreports) and Glascock (https://archaeometry.missouri.edu/murr_database.html), many are not systematically available.

It has been suggested that one could compile data for a region into a large spreadsheet (Lee Drake, personal communication), and the chemistry of source samples will emerge from many analyses of source samples from a given source on reliable instruments. While this approach will work within a research team or trusted group, the spreadsheet approach is cell-focused and lacks metadata, such as lineage tracking for longer term accumulation of data.

In recent years, a number of data science projects [Data Version Control, Open knowledge International, AMI at CERN] use version control, such as Git used in software engineering, for compiling scientific data itself. The principals of code branches that can be merged apply here with version control systems such as Git in that they have the advantage of providing rigorous metadata for tracking edit histories and lineage to a defined data set or table of concentrations, as well as the associated spectra and calibration files. If a particular instrument or analysis episode turns out to have been problematic, the entire branch can be separated and perhaps removed.

An XRF workflow can be partially scripted so that the results of an analysis are checked against a standard with a defined error range per element, and then version control (e.g., *git commit*) updates a repository with spectra and associated calibration files, as well as parsing the concentration values to tables that provide the elemental concentration values per obsidian chemical group (source).

Conclusion

We present an approach here where certain obsidian geochemistry data sets, principally those derived from Neutron Activation Analysis, are shared in an open repository and available for researchers interested in source geochemistry. This chapter describes a platform for using the relatively sensitive NAA geochemistry with measurements from multiple calibrated XRF instruments per source to complement the NAA chemistry.

References

Brunsdon, Chris, and Alexis Comber
  2021   Opening Practice: Supporting Reproducibility and Critical Spatial Data Science. *Journal of Geographical Systems* 23(4): 477–496.

Frahm, Ellery
  2019   Introducing the Peabody-Yale Reference Obsidians (PYRO) Sets: Open-source Calibration and Evaluation Standards for Quantitative X-ray Fluorescence Analysis. *Journal of Archaeological Science: Reports* 27: 101957.

Glascock, Michael D.
  2011   Comparison and Contrast between XRF and NAA: Used for Characterization of Obsidian Sources in Central Mexico. In *X-Ray Fluorescence Spectrometry (XRF) in Geoarchaeology*, edited by M. S. Shackley, pp. 161–192. Springer, New York.
  2020   A Systematic Approach to Geochemical Sourcing of Obsidian Artifacts. *Scientific Culture* 6(2): 35–47.

Glascock, Michael D., Geoffrey E. Braswell, and Robert H. Cobean
  1998   A Systematic Approach to Obsidian Source Characterization. In *Archaeological Obsidian Studies: Method and Theory*, edited by M. Steven Shackley, pp. 15–64. Plenum, New York.

Glascock, Michael D., and J. R. Ferguson
  2012   Report on the Analysis of Obsidian Source Samples by Multiple Analytical Methods. Internal report available from the Archaeometry Laboratory, University of Missouri, Columbia.

Govindaraju, K.
  1994   Compilation of Working Values and Sample Description for 383 Geostandards. *Geostandards Newsletter* 18(S1): 1–158.

Johnson, Lucas R. M., Jeffrey R. Ferguson, Kyle P. Freund, Lee Drake, and Daron Duke
  2021   Evaluating Obsidian Calibration Sets with Portable X-Ray Fluorescence (ED-XRF) Instruments. *Journal of Archaeological Science: Reports* 39: 103126.

Kansa, Eric C., and Sarah Whitcher Kansa
  2021   Digital Data and Data Literacy in Archaeology Now and in the New Decade. *Advances in Archaeological Practice* 9(1): 81–85.

Marwick, Ben

2017   Computational Reproducibility in Archaeological Research: Basic Principles and a Case Study of Their Implementation. *Journal of Archaeological Method and Theory* 24(2): 424–450.

Stark, Philip B.

2018   Before Reproducibility Must Come Preproducibility. *Nature* 557(7706): 613–614.

Authors

Guillermo Acosta-Ochoa. Mexican. BA in Archaeology from ENAH, MA and PhD of Anthropology from UNAM. Full-time researcher at the Institute of Anthropological Research at the National Autonomous University of Mexico (IIA-UNAM) and co-head of the Prehistory and Human Evolution Laboratory. Director of several projects related to the first settlers of America, early agriculture, and the development of social complexity.

Clive Bonsall studied prehistory and archaeology at Sheffield University under Colin Renfrew and Paul Mellars. He held a junior research post at the British Museum from 1974 to 1978, before taking up a full-time teaching post at the University of Edinburgh (1978–2021). He continues to contribute to teaching and research at Edinburgh University where he holds the title of Emeritus Professor of Early Prehistory. His research focuses on the post-glacial hunting-gathering (Mesolithic) and early farming (Neolithic) societies of Britain and continental Europe. He has conducted fieldwork in Britain, Slovenia, and Romania. His publications comprise 9 books and nearly 200 scientific articles, including 9 on obsidian source and artifact characterization in Southeast Europe.

Eder Antonio Borja-Laguna. Mexican. Archaeologist from ENAH. He currently collaborates in the Promeza Muyil Salvage Project and the Archaeoacoustics of the Nuns´ Quadrangle Project at Uxmal, Yucatan. He has participated in various research projects with photogrammetry, geomatics, and XRF analysis at the Prehistory and Evolution Laboratory (IIA-UNAM).

Steven Brandt's research interests lie in Eastern Africa, where his students and colleagues are engaged in projects on Late Pleistocene refugia and the evolution of modern cultural behavior, the establishment of agriculture and nomadic pastoralism, the ethnoarchaeology of lithic production and hide working, and the protection and conservation of Africa's cultural heritage. His current projects include researching the deep-time record at the Mochena Borago rockshelter and more recent household crafting at the pre-Aksumite site of Mezber.

Luis Alberto Coba-Morales. Mexican. Archaeologist from ENAH. He currently collaborates in the Terronera Archaeological Salvage Project at the INAH-Jalisco Center. He has participated in various research projects with a geoarchaeological perspective and landscape transformation at the Prehistory and Evolution Laboratory (IIA-UNAM).

M. Kathleen (Kathy) Davis is a field archaeologist and X-ray fluorescence specialist at Far Western Anthropological Research Group in Davis, California, who has worked on projects in California and the Great Basin since 1994. She began working with X-ray fluorescence to study obsidian artifact distribution in the early 1990s, training at the geosciences laboratory at UC Berkeley and at Spectrace Instruments in Colorado. As member of the obsidian sourcing laboratory at Far Western in Davis, she is currently researching new techniques to analyze small artifacts with portable XRF.

Paola Donato is Associate Professor of Geochemistry and Volcanology in the University of Calabria, Italy. Her research activity focused on magmatological and field aspects of different volcanic areas, in Italy and worldwide (Ustica, Mexico, Greece, Iran, Iceland, Ethiopia) and on tephrostratigraphy of Southern Italy. Moreover, she studies the geochemical and micromorphological features of Italian and Mexican obsidians and the volcanic component in fluvial and coastal sediments. During her research activity, she has acquired advanced skills in the use and application of sophisticated analytical techniques. She is author of several papers published in international journals.

Daron Duke is a company Principal at Far Western Anthropological Research Group and the Director of the Far Western XRF Lab out of Henderson, Nevada. He has worked in the Great Basin and California for more than 25 years, specializing in regional lithic economy and land use changes through time. His work draws heavily from XRF sourcing and obsidian hydration dating studies rooted in cultural resource management projects. He looks to disseminate these data into the published scientific literature when possible and appropriate.

Franco Foresta Martin is a geologist with an MA in Geochemistry from the University of Palermo, Italy. He is currently Associate Researcher at the Istituto Nazionale di Geofisica e Vulcanologia (INGV). His research fields include archeometry, volcanology, and geomorphology. Much of his research and publications are dedicated to the provenance of obsidian artifacts found on the island of Ustica (Sicily, Italy) and the geochemical characterization of the four main obsidian outcrops in Italy (Lipari, Pantelleria, Palmarola, and Monte Arci) exploited during prehistoric times. He is the founder and director of the Ustica Island Museum of Earth Sciences.

Kyle Freund is a Principal Investigator at Far Western Anthropological Research Group, Inc. and an Adjunct Assistant Professor at the University of Nevada, Las Vegas. He is a prehistoric archaeologist and lithic analyst with regional specializations spanning the Old and New Worlds, including island and coastal contexts of the Mediterranean and Florida peninsula, as well as desert environments of the US Great Basin and Arizona. Prior to joining Far Western, Kyle served as an Assistant Professor of Anthropology and Bank Atlantic Endowed Teaching Chair at Indian River State College in Ft. Pierce, Florida.

Kata Furholt works as a postdoctoral researcher and lecturer at Kiel University. She earned her PhD at ELTE in Budapest (Hungary) in 2019 (prior to 2024, her name was Kata Szilágyi). The topic of her research was the chipped stone tool production activity of the southeastern group of the Late Neolithic Lengyel culture. For 10 years, she worked at the Móra Ferenc Museum, Szeged. Dr. Furholt has teaching experience in prehistoric archaeology, stone tool technology, the Neolithic and Copper Age periods of the Carpathian Basin, and museology. She worked as an invited lecturer at the University of Szeged (2015–2019) and as a postdoctoral guest researcher-lecturer at the University of Oslo (2020–2021) and Kiel University (2021–2022). She accumulated extensive fieldwork experience in the Carpathian Basin during her studies and work. Beyond lithic analyses (tool procurement and production activities), she is interested in the Neolithic and Copper Age of the European continent and topics of social systems (social complexity, inequality), cultural anthropology, burial rituals, the concept of value, geoarchaeology, the connection between landscapes and communities, as well as cultural memory.

Víctor Hugo García-Gómez. Mexican. BA in Archaeology from ENAH, and MA of Anthropology from UNAM. His research interests include the exploitation of raw materials in obsidian deposits, as well as the characterization and development of obsidian exchange routes in the prehispanic period.

Michael D. Glascock recently retired as a Research Professor at the University of Missouri Research Reactor (MURR). He was the leader of the Archaeometry Lab at MURR from 1988 until 2023 and co-authored more than 500 articles and 10 books on analyses of archaeological materials. He received the Rould Fryxell Award (2009) for contributions to archaeology through interdisciplinary research from the Society for American Archaeology and the Pomerance Award (2011) for contributions to archaeology from the Archaeological Institute of America.

Ferenc Horváth studied archaeology and history at Szeged University (Hungary), where he obtained the post of Assistant Professor (1973–1975), before securing a full-time research position at the Móra Ferenc Museum, Szeged (1976–2011). Since 1997, he has continued half-time teaching as Associate Professor, after obtaining his CSc (PhD) degree at the Hungarian Academy of Sciences. From 2006 to 2011, he was the Director and Deputy Director of the Szeged Museum. His research has focused on the Neolithic of Southeast Europe, 14C chronology, and polished stone tools. His publications comprise one book and nearly one hundred scientific articles. He directed the excavation of the tell site at Gorzsa from 1978 to 1996.

Richard E. Hughes is the Director of the Geochemical Research Laboratory, Sacramento, California, and a Research Associate at the Archaeological Research Facility. He has conducted geochemical provenance analyses and studies of prehistoric obsidian conveyance since 1978, focusing primarily on California and the Great Basin of North America.

Lucas R. M. Johnson is a Senior Laboratory Scientist at Far Western Anthropological Research Group, Inc. He specializes in lithic technology, XRF, and the study of craft production in North America and Mesoamerica.

Eric C. Kansa is an archaeologist with a research focus on data management in cultural heritage domains. He is the architect behind Open Context, an open access platform for archaeological data dissemination. As Technology Director for Open Context, he works to develop and implement good practices in informatics, ethics, and research data management.

Enrico Massaro is a retired Professor of Physics and Astrophysics at the Sapienza University of Roma and is currently working at the National Institute of Astrophysics in Rome, Italy. His main field of research is High Energy Astrophysics of galactic and extragalactic sources, and other interests are the history, epistemology, and the cultural and social impact of science. He is author/coauthor of more than 350 scientific papers.

Emiliano Ricardo Melgar Tísoc. Mexican. BA in Archaeology from ENAH, MA and PhD of Anthropology from UNAM. Full-time researcher at the Museum of the Great Temple. Coordinator of the project Style and technology of lapidary objects in Ancient Mexico. His research topics include lapidary technology, commerce, tribute, prehispanic navigation, workshops, ancient mines, archaeometry, and traceology.

Donna Nash is Associate Professor of Anthropology at the University of North Carolina, Greensboro. She earned her PhD at the University of Florida and was a postdoctoral associate for the *Ancient Americas* permanent exhibit at the Field Museum, Chicago. Her field research examines imperial expansion from the perspectives of state agents and members of subject groups using data gathered through household archaeology. Her approach directly addresses the methodological problems arising from studying "prehistoric" empires and their material culture. She seeks to relate households to the state and define institutions through architecture and activities that were essential to imperial governance. Dr. Nash has published on a number of themes that intersect with these goals, including on feasting in *Wari: Lords of the Ancient Andes* (Bergh 2012), the built environment in *Vernacular Architecture of the Pre-Columbian Americas* (Halperin and Schwartz 2017), craft production in *Journal of Anthropological Research* (2019), and ritual in *Journal of Anthropological Archaeology* (Nash and deFrance 2019).

M. Steven Shackley is a Professor Emeritus of Anthropology at the University of California, Berkeley, and former Director of their NSF-funded Geoarchaeological XRF Laboratory. Although retired, he continues to teach as an adjunct at the University of New Mexico and serves as Director of a new Geoarchaeological XRF Laboratory in Albuquerque. Dr. Shackley is a highly respected archaeological scientist and recipient of the 2019 SAA Fryxell Award for Interdisciplinary Research. His 30-plus-year career has spanned the fields of anthropology, archaeology, geology, and museum studies. Dr. Shackley is the author of over 150 journal articles and book chapters, as well as numerous technical reports. He has also published several prominent monographs and edited volumes, including *X-Ray Fluorescence Spectrometry in Geoarchaeology* (2011), *Obsidian: Geology and Archaeology in the North American Southwest* (2005), and *Archaeological Obsidian Studies: Method and Theory* (1998). He is perhaps best known for his work on obsidian characterization in the American Southwest, where he has been on the forefront of elemental characterization studies using XRF spectrometry.

Benjamin D. Smith's research explores lifeways and material culture of hunter-gatherers. His current projects are aimed at identifying obsidian sources and examining the organization of lithic technology around major environmental, evolutionary, and social changes in the Pleistocene Horn of Africa. His work in southern Ethiopia explores the natural and artifactual stone landscapes visited by hunting and gathering peoples for the past two million years. He recently received a postdoctoral study grant from the Fyssen Foundation for a project entitled "Living and making on landscapes of stone: understanding *Homo sapiens*dispersal and the emergence of the human mind through obsidian sourcing in the Late Pleistocene Horn of Africa'. This project will be based out of the Cultures et Environnement, Préhistoire, Antiquité, Moyen Âge (CEPAM) laboratory at Côte d'Azur University in Nice, France.

Reyna Beatriz Solís-Ciriaco. Mexican. BA in Archaeology from ENAH, MA and PhD of Anthropology from UNAM. Researcher of the project "Style and technology of lapidary objects in Ancient Mexico." Her research topics include artisans, specialized production, style, tradition, symbolism, technological choices, archaeometry, and experimental archaeology.

Elisabetta Starnini studied prehistory and archaeology in Italy, at Genova University, where she obtained her PhD in Prehistory of Mediterranean Countries. Following a tenure-track researcher position in Prehistory, she now holds a full-time teaching post as Associate Professor of Prehistory and Protohistory at the University of Pisa (from 2022). Her research focuses on lithic raw material sourcing, the Mesolithic/Neolithic transition, the Neolithic of the Carpathian Basin, and Italian and Balkan prehistory. She is involved in the study of the lithic assemblages from the excavations at Gorzsa (HU). She has conducted fieldwork in Italy, Hungary, Pakistan, Oman, and Greece. Her publications comprise 6 books and more than 250 scientific articles, including 2 recently published on obsidian characterization of Neolithic artifacts from an Italian cave site.

Marcin Szeliga is an Assistant Professor at the Institute of Archaeology of the Maria Curie-Skłodowska University in Lublin. His previous scientific interests focused on the Early Neolithic period on the northern side of the Carpathians and especially on the technology of processing and distribution of various flint raw materials and obsidian in this period.

Nicholas Tripcevich is the Associate Director of the Archaeological Research Facility at UC Berkeley. He is an archaeologist with expertise in geospatial approaches and obsidian studies in South America and the US West.

Robert H. Tykot is a Professor in the Department of Anthropology at the University of South Florida in Tampa. He has conducted obsidian studies for more than 35 years on 5 continents – mostly in the Mediterranean, where he has done extensive survey of island geological sources and used a variety of analytical methods. He has more than 60 formal publications on his obsidian research. His website may be found at http://faculty.cas.usf.edu/rtykot/.

Barbara Voytek received her MA from Harvard University in 1978 and her PhD from the University of California at Berkeley in 1985. The title of her dissertation is "Lithic Exploitation in Neolithic Southeast Europe." Subsequently, she was the Executive Director of the Center for Russian and East European Studies at Stanford University. In 1989, she became the Executive Director at the Institute for Slavic, East European, and Eurasian Studies at the University of California at Berkeley. She spent twenty years as a student and then as a research scholar excavating Neolithic sites in the former Yugoslavia. In 1991, she began research at the Grotta dell'Edera in the karst of Italy near Trieste. That project concluded in 2001. Dr. Voytek is currently a Research Associate at the Archaeological Research Facility of the Department of Anthropology at the University of California at Berkeley. Her publications include several scientific articles on lithic assemblages and use-wear analysis from the most important European Neolithic sites.

Dagmara H. Werra is an Assistant Professor at the Institute of Archaeology and Ethnology of the Polish Academy of Sciences in Warsaw. In her professional career, she has focused on prehistoric flint mining, the use of flint in the Stone Age and Metal Ages, and the identification and use of siliceous rocks by prehistoric communities.

INDEX